WATER PRODUCTIVITY OF SUNFLOWER UNDER DIFFERENT IRRIGATION REGIMES ON GEZIRA CLAY SOIL, SUDAN

Eman Rahamtalla Ahmed Elsheikh

Thesis committee

Promotor

Prof. Dr E. Schultz
Emeritus Professor of Land and Water Development
UNESCO-IHE Institute for Water Education
Delft, the Netherlands

Co-promotors

Prof. Dr Hussein S. Adam
Professor of Water Management and irrigation Institute
Gezira University
WadMedani, Sudan

Dr Abraham Mehari Haile
Senior Lecturer in Hydraulic Engineering
UNESCO-IHE Institute for Water Education
Delft, the Netherlands

Other members

Prof. Dr P.J.G.J. Hellegers, Wageningen University
Dr Abdel Hadi AW Mohamed, Arabian Gulf University, Kingdom of Bahrain
Dr Eyasu Yazew Hagos, Mekelle University, Ethiopia
Dr F.W.M. van Steenbergen, MetaMeta Research, 's Hertogenbosch

This research was conducted under the auspices of the SENSE Research School for Socio-Economic and Natural Sciences of the Environment

WATER PRODUCTIVITY OF SUNFLOWER UNDER DIFFERENT IRRIGATION REGIMES ON GEZIRA CLAY SOIL, SUDAN

Thesis

submitted in fulfilment of the requirements of
the Academic Board of Wageningen University and
the Academic Board of the UNESCO-IHE Institute for Water Education
for the Degree of doctor
to be defended in public
on Thursday, 8 October, 2015 at 2:30 p.m.
in Delft, the Netherlands

by

Eman Rahamtalla Ahmed Elsheikh
Born in Khartoum, Sudan

CRC Press
Taylor & Francis Group
Boca Raton London New York

CRC Press is an imprint of the
Taylor & Francis Group, an **informa** business
A BALKEMA BOOK

First issued in hardback 2018

CRC Press/Balkema is an imprint of the Taylor & Francis Group, an informa business

© 2015, Eman Rahamtalla Ahmed Elsheikh

Published by:
CRC Press/Balkema
PO Box 11320, 2301 EH Leiden, The Netherlands
e-mail: Pub.NL@taylorandfrancis.com
www.crcpress.com – www.taylorandfrancis.com

www.crcpress.com - www.taylorandfrancis.co.uk - www.balkema.nl

ISBN 13: 978-1-138-37339-6 (hbk)
ISBN 13: 978-1-138-02914-9 (pbk)

Table of contents

Acknowledgements

I would like to express my sincerest gratitude to my supervisor, Prof. E. Schultz who has supported me throughout my research with his patience and knowledge. I would like to thank him for encouraging and supporting me and his brilliant comments and suggestions. I attribute the level of my PhD degree to his encouragement and effort. Without him this thesis, too, would not have been completed or written.

I would especially like to thank my mentor Dr. Abraham Mehari Haile for his advices during the fieldwork and research phase and for his excellent guidance. He has been invaluable on both an academic and a personal level, for which I am extremely grateful.

A special thank to my local supervisor Prof. Hussein Adam, who always supported and encouraged me during my study. He gave me constructive comments and warm encouragement. He looked after me as my real father.

My sincere gratitude goes to Prof. Adil Omer Salih, the deputy director of the Agricultural Research Corporation, for his help and giving me the chance to continue my PhD study in the Netherlands.

Many thanks go to Prof. Kamal Alsiddig, the Head of the Research Program and International Collaboration, Agricultural Research Corporation, for his support and giving me the opportunity to do this PhD study. Great appreciation is also given to the Head of the Forestry Research Centre, Prof. Mohamed ElMoukhtar, for his support and giving me freedom to conduct my fieldwork in Gezira Research Station, and for his valuable comments and knowledge.

I would also like to thank the Head of Gezira Research Station, who gave me permission to conduct all the fieldwork in the Gezira Research Station Farm during the three years of my fieldwork.

I would like to acknowledge the Land and Water Research Centre for their support and assistance from the start of this study. Special thanks to the former Head of the Centre Prof. Abdelah Alhajwa and the present Head Prof. Mohamed Ahmed Salih.

My sincere gratefulness goes to Dr. Abdel Hadi Abdel Wahab for his good advices and guidance during my PhD study. I learnt from him much science, skills and knowledge.

I gratefully acknowledge Dr. Imad Ahmed Ali for his guidance and useful comments and suggestions. Also, my gratefulness goes to Dr. Ihsan Mustafa for her help and willingness to advise me during my fieldwork. I appreciate their assistance.

I would like to thank Mr. Yasir Elamein the Head of the Agricultural Machinery Section for his support during the preparation of the field for all the three seasons. Really he was a good and helpful person.

I am particularly grateful for the assistance given by Hala and Abdallah my best comrades. They were always willing to help me during the fieldwork. Many thanks go to my colleagues (ladies) in the guest house of the Agricultural Research Corporation for their helping and caring.

Also many thanks go to my friends in UNESCO-IHE, Hassana Moustapha from Nigeria, and Yasir, Khalid, Agmed, Ishraga, Reem, Eiman, Sara, the Sudanese friends for their support and giving me advices, assistance and sharing their knowledge and experiences.

I would like to offer my special thanks to the Water Management Research Section staff for their assistance and support to collect data, as well as to the staff of the Gezira Research Station for their collaboration and to offer all the facilities to conduct this study in the farm.

I would like to express my gratitude to Dr. Mergani Saeid for his supporting and insightful comments and suggestions.

I extend my special thanks to my colleague Ahmed from whom I have received consistent warm support and advices.

The research for this thesis was undertaken with the financial assistance of the Netherlands Fellowship Programme (NFP). Without it my dream could not become true. Deep thanks for their support.

My sincere gratitude goes to the UNESCO-IHE employees, especially Jolanda for her support and assistance during all the time of my study. She was patient with our continuous requests.

Thanks to all my colleagues and friends in WadMedani and Khartoum whom I have not mentioned for their caring and support and for praying for me to finish this hard work on my PhD study.

Finally, I want to send my deep love and thank to my family, my mother, my beloved sisters and my brother. Their generous and unselfish love always warms my heart. I thank them for kindly supporting me to convert my dream into reality.

To the souls of my Father, my beloved sister Amira, and to Prof. Eltahir Elsiddig, really we miss them.

Summary

Due to the growing world population, agricultural production will need to increase. Yet the portion of fresh water currently available for agriculture (72%) is decreasing. Under current water scarcity conditions, the limited available water would have to be used more efficiently. Previously agricultural research has focused primarily on maximizing production. In recent years, the researches focused on the production system, notably the availability of either land or water, where water is the limiting factor in crop cultivation. In regions where water scarcity is the principle limiting factor for cultivation, farmers are interested in growing crops that are able to adapt to drought conditions.

The Gezira Scheme is Sudan's oldest and largest gravity irrigation system. The total area of the Gezira Scheme in the central clay plains is about 880,000 ha, located between the Blue Nile and White Nile rivers. The scheme has played an important role in the economic development of the country, and is a major source of foreign exchange. It has also contributed to the national food security and in generating a livelihood for the 2.7 million people who now live in the scheme.

In the scheme the farmers do not own the land, they are tenants. The area is divided among 120,000 tenancies with an average area of about 7.3 ha. The original size of a regular tenancy in the old Gezira Scheme was 16.8 ha, but a number of tenancies have been converted to half-tenancies over the years. Tenant participation in agriculture has fallen over two generations of rising living standards and better education. It is estimated that over half of all tenants are now no longer involved in farming and most of those who are still involved, have an outside job as well.

The Gezira Scheme was managed by the Sudan Gezira Board (SGB) as a government enterprise. The SGB used to work closely with the Ministry of Irrigation and Water Resources (MOIWR) that was responsible for the operation, maintenance and management of the main, major and minor canals (irrigation network). The SGB was responsible for the agricultural operation. It had irrigation committees along the minor canals and representatives of each of these committees constituted the Irrigation Committee at the block level. In 2005 the new act for the scheme management was established by the Sudanese government "the Gezira Act 2005". The responsibility of the operation and maintenance of the major and minor canals was assigned to the SGB. Based on this act the SGB established 17,000 water users groups, which are called Water Users Associations (WUA). The responsibilities of the WUAs concern operation and maintenance of the lower level canals and involvement in the operation and maintenance of the minor canals.

In Gezira Scheme most of the farmers in some areas use the short rainy season from June to September for growing crops such as Sorghum and vegetables. The farming system of the Gezira Scheme is dominated by crop production. The major crops grown are Sorghum, Wheat, Groundnut and the oilseed crop Sesame. The cropping calendar is as follows: from June to July is the time for Sorghum, Cotton, Groundnut, Sesame and vegetables by using crop rotation and from November to December for growing Wheat and vegetables.

Sunflower (*Helianthus annuus L.*) is an important oil crop in the world as well as in Sudan. It is a new edible oil crop in Sudan, the seeds have an oil content of 40 - 50% and 30% digestible protein and can thus be used as a source of food for humans, or as poultry feed. Sunflower cake can be used as animal feed. It can be cultivated as winter crop under irrigated conditions and as summer crop under rainfed conditions. Sunflower has become an important oil crop for both farmers and consumers in Sudan. It is a crop that fits well in the local cropping system and is considered as the most important oil crop of the country. Although experiments on Sunflower in Sudan started as early as the 1980s, commercial production started late and focused mainly on rainfed cultivation. The cultivated areas in 1990 - 1991 reached 123,000 ha with an average yield of 714 kg/ha. Recently, high yielding hybrids were introduced and grown under irrigation. In Sudan Sunflower is now grown in two seasons (winter and summer) and is recognized as a crop with high potential that can successfully meet future oil requirements. In Gezira Scheme the cultivated area with Sunflower for the summer and winter seasons of 2013 were 2570 and 619 ha respectively. While the average yield obtained from farmers fields ranged between 1.2 and 2.2 t/ha (0.5 - 0.9 t/fed).

Sunflower is categorized as a low to medium drought sensitive crop. The drought-tolerant nature can be attributed to its extensive root system, which can extract water and nutrients to a depth of 3 m. The available information shows that seed yield response to water is usually greatest when Sunflower is irrigated at a late flower-budding stage before the flowering stages.

Knowledge of the effects of irrigation scheduling on Sunflower production and water productivity under water stress conditions is becoming increasingly important. Irrigation scheduling is particularly important since many field crops are more sensitive to water deficit at specific phenological stages. In crop production instead of achieving maximum yield from a unit by full irrigation, water productivity can be optimized within the concept of deficit irrigation. Deficit irrigation is an optimal strategy in which irrigation is applied during drought-sensitive growth stages of a crop. Deficit irrigation has been widely investigated as a valuable and sustainable production strategy in dry regions. By limiting water applications to drought sensitive growth stages, this practice aims to maximize water productivity and to stabilize rather than maximize yield. In addition correct application of deficit irrigation requires a thorough assessment of the economic impact of the yield reduction caused by drought stress.

Sunflower has several growth stages: emergence, vegetative, reproductive, flowering, seed formation and maturity. Water stress in each stage results in reduction in seed yield and oil content. The treatments, which were conducted to study the effect of water stress at different growth stages, showed that Sunflower was significantly affected by water stress that occurred in the sensitive flowering, seed formation and ripening stages. The highest yield was obtained from the treatments that were not stressed during these stages. It was also found that deficit irrigation at the early flowering stage reduced seed yield by 25% by reducing both seed number and individual seed weight, while at the mid flowering stage seed yield was less affected by deficit irrigation. It was also reported that Sunflower needs to be irrigated three times with full or limited irrigation water at heading, flowering and seed filling stages for high seed yield.

Sunflower is a short duration crop and can be fitted well in our present cropping system without bearing any major change in agriculture setup. The planting pattern plays a key role in increasing the yield of Sunflower. Flowering is the most critical stage to water stress caused by deficit irrigation. For this reason limited irrigation in early and mid flowering stages needs to be avoided, while deficit irrigation at seed formation can be acceptable whenever irrigation water is limited. Previous studies showed that the water requirements of Sunflower vary from 600 to 1000 mm depending on the climate and length of total growing period. High evapotranspiration rates occur during the seed setting and early ripening stage. The water requirements of Sunflower are relatively high compared to other crops. Nevertheless it has the ability to withhold short periods of soil moisture deficit up to 15 atmosphere tension.

Water use of a crop with satisfactory available soil moisture supply is primarily affected by its canopy and weather conditions. These effects are represented by the seasonal crop coefficient and the reference evapotranspiration (ETo) of the atmosphere. The crop coefficient indicates the fraction of potential evapotranspiration (ETp) that the crop is expected to utilize on a given day. The crop coefficient value typically changes with the growing stage.

Research on the effect of water stress on Sunflower yield and water productivity is scarce in Gezira Scheme. In this study determination of the crop water requirements and crop coefficients was the main objective. The study was carried out at the Gezira Research Station Farm to determine the crop water requirement and water productivity and to investigate the effect of deficit irrigation on Sunflower yield and yield components under Gezira clay soil conditions. Three experiments were conducted to test Sunflower seed yield under different irrigation intervals and two with different intra-row plant spacing for winter and summer seasons. The first experiment was conducted to determine the crop coefficient (Kc) of Sunflower and the effect of different irrigation intervals on Sunflower yield and yield components in the winter season, and also to investigate the effect of water stress at specific growth stages. Three irrigation intervals were selected to investigate the sensitivity of two growth stages (flowering and seed filling) to water stress. Irrigation intervals of 10, 15 and 20 days were tested on the crop stage of 50% flowering and on the seed filling stage, which were compared to full irrigation (weekly irrigation). Sunflower Hysun 33 (Hybrid) was selected to evaluate its water productivity and economic water productivity under seven different irrigation treatments. The evaluation was done based on the seed yield obtained from the three growing seasons: the first season started on the 14th of November 2011, the second season on the 19th of November 2012, while the third season started on the 5th of December 2013. It was concluded that the highest yield was obtained under weekly irrigation and 10 days intervals after the flowering and seed filling stages. The 20 days irrigation intervals decreased the yield with more than 40%.

Based on the extensive field data collection the crop coefficient of Sunflower has been determined by using the procedure recommended by the Food and Agriculture Organization of the United Nations (FAO). This procedure is based on a calculation of the soil moisture content for estimating evapotranspiration. From the meteorological data obtained from the Meteorological Station located nearest to our field, the reference evapotranspiration (ETo) was calculated by using the FAO EToCalc software programme (2009). The results revealed that the crop coefficients for the initial stage, development stage, mid season and late season stages were 0.53, 1.20, 1.30 and 0.63 respectively.

Under weekly irrigation a maximum seed yield of respectively 3130, 3140 and 3100 kg/ha was obtained for the first, second and third year. Irrigation at 10, 15 and 20 days intervals after the flowering stage reduced the seed yield by respectively 15, 23 and 34% in the first season, while in the third season it was 8, 20 and 31%. Moreover, the yield reduction in irrigation treatments of 10, 15 and 20 days after seed formation was 10, 25 and 30% and 9, 24 and 26% in the first and second season respectively. Highest water productivity was achieved under weekly irrigation and ranged from 0.32 to 0.36 kg/m^3. The lowest one was obtained from 20 days irrigation after the flowering stage (0.21 and 0.26 kg/m^3). Water deficit during the seed formation stage reduced the seed yield as compared with full irrigation, but the reduction was much less than when stress occurred during the flowering stage.

Net incomes from each irrigation treatment were studied to evaluate the effect of water deficit on seed yield economically. Net income depends mainly on production costs. However, 8 - 9% higher inputs in production cost for 10 days irrigation intervals compared to irrigation intervals of 15 days resulted in a higher net income of US$721 /ha and US$ 866/ha in the first and second season respectively. This revealed that the gain from higher yield is greater than the associated higher irrigation cost for this treatment.

The second and third experiments were conducted in the winter and summer seasons respectively to evaluate seed yield obtained from irrigated Sunflower (10, 15 and 20 days irrigation intervals) under two intra-row plant spacings (30 and 40 cm) and tested with one cultivar Hysun 33 (Hybrid) in the winter season 2012/2013. The measurements of crop parameters such as plant height, head diameter, total seed yield and 100-seeds weight were recorded. The results of the winter experiments revealed that the highest seed yield was obtained from an irrigation interval of 10 days with 40 cm intra-row plant spacing (3290 kg/ha) followed by the 15 days interval irrespective of the intra-row plant spacing (3120 and 3050 kg/ha) in the first and second season respectively. The lowest seed yield of respectively 1890 and 1830 kg/ha was obtained from 20 days irrigation interval with both intra-row plant spacings. The highest water productivity was achieved from an irrigation interval of 15 days of 0.42 - 0.45 kg/m^3 and the lowest water productivities were obtained from irrigation at 20 days interval of 0.31 - 0.32 kg/m^3 under both intra-row plant spacings. The highest economic water productivity of US$ 0.29/m^3 was obtained from an irrigation interval of 15 days in the first season and the lowest economic water productivity of US$ 0.17/m^3 was obtained from an irrigation interval of 20 days in the first season. This shows as well that the economic indicator may be an appropriate tool for assessing impacts of deficit irrigation and water prices.

The summer experiments were conducted to study the effect of different irrigation intervals compared to seed yield under rainfed conditions. Two crop varieties, Hysun 33 (Hybrid) and Bohooth-1 (local variety), were grown starting mid July 2012 and 2013 respectively. Results showed that stressing the crop up to 20 days irrigation intervals reduced seed yield between 40 and 44% for Hysun 33 and between 38 and 44% for Bohooth-1 in the first and second season respectively. While increasing irrigation interval to 15 days did not result in a significant reduction in seed yield (16 - 31%). The results revealed that the differences between 10 and 15 days irrigation interval in seed yield were not highly significant. No significant differences were found between the two varieties in seed yield and water productivity under different irrigation treatments and intra-row plant spacing in the summer season. Compared to the 10 days interval the average amount of water applied was increased by 6% and 11% in the 15 days and 20 days irrigation intervals respectively. The contribution of rain in water supply was respectively 18% for the 10 days, 24% for the 15 days and 24% for the 20 days interval. Only a part of this rain can be used to satisfy the crop water demand at the establishment stage. At this end, comparison of the results in seasonal water applied between the three irrigation treatments indicate that the 10 and 20 days intervals were recorded higher and lower in irrigation water application respectively.

Higher water productivity was obtained from intervals of 10 days (0.41- 0.42 kg/m^3) and 15 days (0.39 - 0.44 kg/m^3) for Hysun 33 in the first and second season, while the highest water productivity (0.41 kg/m^3) was obtained from the irrigation interval of 10 days irrespective of the intra-row plant spacing for Bohooth-1 in the first season and 0.39 kg/m^3 was obtained from the same irrigation interval in the second season. However, there were significant differences recorded between the two treatments.

Benefit-cost ratios (B/C) of all full irrigation treatments were higher than 1. Therefore, we can conclude that Sunflower cultivation in Gezira Scheme with providing full irrigation at 10 days and/or 15 days intervals is an economically viable option at current market prices. The total production costs were higher under full irrigation treatment compared with the other treatments, while the benefit was more or less acceptable under the prevailing economic situations.

Crop growth simulation models have been developed for predicting the effect of water and soil on biomass, grain yield and water productivity of different crops. In this study the water driven crop model AquaCrop, developed by FAO, was calibrated and validated for Sunflower under different irrigation regimes. Calibration was done by using obtained data from the experimental field of 2011 and validation with the data of 2012 and 2013. The calibrated and measured data were compared to assess the model performance. Agreement between the calibrated and measured values was quantitatively evaluated by using the root mean square error (RMSE) and the index of agreement (d). The lower the RMSE and the higher d are the better the model performance is.

Statistical indicators RMSE and d suggest that the model can be used to highly reliably assess the yield and canopy cover under full and deficit irrigation conditions. In this study the AquaCrop model predicted seed yield with RMSE 0.05 - 0.16 t/ha, d 0.87 - 0.98 and canopy cover with RMSE 1.9 - 10.1% and d 0.99 more precisely compared to simulated water productivity (RMSE 0.10 - 0.14 kg/m^3 and d 0.27 - 0.64). There was a tendency to over-estimate water productivity for all irrigation treatments for the winter season. The AquaCrop model followed the same trend in predicting seed yield and water productivity under summer conditions. The seed yields were predicted with RMSE of 0.01 - 0.12 t/ha and d 0.89 - 0.99 and the difference between predicted and measured values was 3.1 - 16.7%. The model over-estimated the water productivity with RMSE 0.01 - 0.03 kg/m^3 and d 0.31 - 0.88 while the deviation was 9.8 - 37.5% of all irrigation treatments.

Based on these results application of the AquaCrop model is recommended under conditions of limited data and yield predictions under different water supply. Accordingly, based on the results of the study it is recommended to conduct research programs to adopt deficit irrigation as a new irrigation strategy with recommended agronomic practices that would increase yield and water productivity under Gezira conditions. In addition, it is recommended to consider optimum water use applications for Sunflower that maximize yield and returns on irrigation water under different climatic conditions and changing market prices.

Deficit irrigation has been conducted in different field trials to assess its effect on Sunflower yield and water productivity. Results showed that the yields obtained under different irrigation levels were higher compared to farmers yield even under severe water stress. Moreover, water productivity was higher when the crop received less water compared to full irrigation. Thus, these results confirm that irrigation at 15 days interval (mild water stress) would be acceptable for improving water productivity when there is water shortage during the growing season. When land is not a limiting factor, intra-row plant spacing of 40 cm would have to be practiced with irrigation at 10 days interval for maximizing yield and to gain more profit.

1   Introduction

The world population of 7 billion is expected to increase with 35% and to reach about 9.3 billion by 2050 (United Nations, Department of Economic and Social Affairs, Population Division, 2011). The increasing population will result in considerable demand for food. Additionally, water demand from non-agricultural sectors will keep increasing in both developed and developing countries. About 40% of the land in the world is under arid and semi-arid climatic conditions (Elias, 2001). Efficient use of rainwater and optimization of water productivity (WP) are important in such conditions. In the dry areas water is the most limiting resource for improving agricultural production. Irrigated agriculture is a major consumer of fresh water, over 45% of global food is produced by irrigated agriculture. In the future, however, the necessary growth in food production has to be achieved with less water.

Sudan has the largest irrigated area in Sub-Saharan Africa and the third largest in Africa with 1.3 million ha at present irrigated from the Blue Nile. The crop production in the irrigated agriculture sector is low and fluctuating due to low production prices, lack of foreign currency and import rules, which have limited the availability of essential production inputs and spare parts (International Fund for Agricultural Development (IFAD), 1992). In the agricultural production the irrigation sector plays a very important role. Although the irrigated area constitutes only about 11% of the total cultivated land in Sudan, it contributes more than half of the total volume of the agricultural production. Irrigated agriculture has become more and more important over the past few decades as a result of drought and rainfall variability and uncertainty. It remains a central option to boost the economy in general and to increase the living standard of the majority of the population. Sudan is generally self-sufficient in basic foods, albeit with important inter-annual and geographical variations, and with wider regional and household disparities in food security prevailing across the country.

The agricultural sector is the most important economic sector in Sudan. It created 39% of the gross domestic product (GDP), employed about 80% of the population, and contributed 80% of the country's exports in the late 1990s. Most of the agricultural activities are concentrated near the Blue Nile River and Nile River. The Nile River starts at Khartoum where the Blue Nile River and the White Nile River merge.

The Al Gezira Irrigation Scheme is located between the White Nile and the Blue Nile rivers upstream of Khartoum. It is the most important agriculture scheme in Sudan. It is also one of the largest irrigation schemes in the world, the largest in Africa as well as in Sudan. The scheme area is about 880,000 ha, located in WadMedani City area south of Khartoum. At present Sorghum, Wheat and Groundnuts are being cultivated instead of Cotton in an effort to maintain Sudan's self-sufficiency in foodstuffs.

Knowledge of crop water requirements (CWR) is decisive for successful water management of irrigated agriculture schemes. The calculation of the CWR is dependent on the prevailing environmental conditions and on the reference evapotranspiration (ETo). Previous studies assumed that the irrigated area would be 3.15 million ha, which is less than 4% of the countries arable land (Abdelhadi et al., 2000). This calls the attention for very efficient use of irrigation water. In the seasons where there is insufficient water for the crop demand, optimum use of irrigation water is essential for water resources management. Optimum use implies efficient irrigation water use and proper timing of irrigation so as to conform to critical stages of growth of the crop concerned. It is therefore important to analyze the effect of water supply on crop yield. So if the water supply does not match the crop water requirements, the actual evapotranspiration (ETa) will be below the potential evapotranspiration (ETp). This means that water stress may occur in the plants. The effect of water stress on growth and yield depends on the crop varieties, and on the magnitude and time of occurrence of water deficit. Hence, in order to optimize the crop water needs for crop production, thorough knowledge of ETa, CWR, critical crop growth stages and irrigation schedules for maximizing production is highly desirable.

Sunflower (*Helianthus annuus L.*) originates from Central and North America. It is considered the fourth oilseed crop in the world, and has a wide range of adaptability and highest oil seed content. Sunflower oil is highly demanded for human consumption and chemical cosmetic industries. The total seed production was estimated at 21 million tons out of 18 million ha (Food and Agriculture Organisation of the United Nations (FAO), 2008). Sunflower grows in different climatic zones ranging from arid under irrigation to temperate under rainfed conditions. The air temperature for best growth ranges between 18 and 25 °C. The total growing period varies and depends on the climate in the different regions. In the subtropics under irrigation, the total growing period is about 130 days. In temperate climates the optimum sowing date for early as well as late maturing varieties is between late winter and early summer.

Sunflower is mainly grown in a wide range of soils under rainfed conditions. Under erratic and low rainfall, a rather deep soil with good water holding capacity is required. It has a deep root system (2 to 3 m), thus, soil water can be extracted up to great depths. Optimum soil pH ranges between 6.0 and 7.5, but at lower values liming may be needful. Fertilizer application is in general 50 to 100 kg/ha N, 20 to 45 kg/ha P, and 60 to 125 kg/ha K. The crop is particularly sensitive to boron deficiency (Allen, 1998).

Sunflower (*Helianthus annuus L.*) was introduced in Sudan in the 20th century, but commercial production at large-scale started in the late 1980s by the private sector. Usually the planting seed is hybrid (FAO, 2011). In recent years the Sunflower cultivation area has increased in Sudan, because of the moderate cultivation requirements and high oil yield. The yield of crops that are cultivated under rainfed conditions has progressively declined with time, owing to changes in both quantity and distribution of rainfall. There is a great need to increase crop production to meet the demand of the rapidly growing population. Smallholder rainfed agriculture has its share in the crop production of Sudan, but in this way crop production cannot significantly be increased. The importance of irrigation to enhance crop production in the water scarce conditions of Sudan is yet to be translated into sustainable development options for irrigated agriculture, which can potentially raise the yield of Sunflower to 5 t/ha.

There is an urgent need to develop and use new approaches and tools, to develop and validate improved models for enhanced crop and water productivity, and to optimize the use of inputs. Improving water productivity requires that more value be obtained from the water that will be used for the crops. With respect to this it is believed that an integrated approach to soil, water and crop is essential for increasing crop productivities. The effect of water stress depends on timing, duration and magnitude of it. Identification of the critical plant growth stages and scheduling of irrigation according to plant's demand is the key factor for conserving water, improving irrigation efficiency and sustainability of irrigated agriculture.

This research concerns options of deficit irrigation for Sunflower cultivation in Gezira Scheme. The water supply from Sennar Dam takes place through a network of irrigation canals. The scheme is managed by the Sudan Gezira Board and farmers have responsibilities for water distribution and canal maintenance through water users associations (WUAs). This study aimed at examining different irrigation water intervals with different agronomical practices to identify the optimum water supply for seed yield and water productivity. Different experimental field trials under various climatic conditions were conducted. The thesis provides reviews of the findings about the water stress and water productivity. Moreover, a new crop modelling approach was used. The research methodology followed in this study is primarily based on extensive field data collection at the Gezira Research Station Farm. The methodology included:

- experimental field trials under winter conditions for three irrigation intervals: 10, 15 and 20 days during the two crop stages of flowering and seed filling;
- measurement of water supply with a current meter at each plot of the experiment to determine the amount of water applied during the whole growing season;
- collect soil moisture samples by the gravimetric method (Augers) from the main plots to estimate the crop evapotranspiration and crop coefficient (Kc) of Sunflower crop;
- experimental field trials under summer conditions (supplementary irrigation with rainfall) with three irrigation intervals and two agronomic practices for optimal seed yield and water productivity;
- experimental field trials under winter conditions with the same three irrigation intervals (easier irrigation scheduling for farmers in Gezira Scheme) and two agronomic practices of 30 and 40 cm intra-row plant spacing to assess the season that would obtain higher seed yield and improve water productivity;
- modelling Sunflower seed yielld and water productivity by using the AquaCrop model and to evaluate the model performance (calibration and validation) with different irrigation water supply levels.

1.1   Structure of the thesis

The introduction (Chapter 1) opens the main research topic of this study through a short description of agricultural production in Sudan and constraints to develop and to improve water productivity under limited water supplies and stress. The Chapter presents the overall needs to develop tools like crop models for enhanced crop and water productivity and for optimizing the use of the inputs.

The overview of the country characteristics in Chapter 2 includes climate, topography, water resources and water use in Sudan. The Chapter further elaborates on the land use and land cover in Sudan. The Chapter ends with describing the problems, the hypothesis and the research questions that had to be answered.

Special consideration is given to the agricultural systems in Sudan. Chapter 3 presents the agriculture sector in Sudan, which is divided into three sectors: mechanized rainfed, traditional agriculture and irrigated agriculture. The Chapter ends with an elaboration of agricultural policies and strategies in Sudan.

A summary is presented of irrigation systems in Sudan. In Chapter 4 the different irrigation systems are briefly explained with a focus on Gezira Scheme as one of the oldest irrigation systems in Sudan. Furthermore, the Chapter presents the other irrigation systems with a short summary.

Chapter 5 gives a brief explanation of the calculation and methodology to estimate the crop coefficients for Sunflower under Gezira conditions. The crop coefficients for the target crop Sunflower were estimated by using the FAO methodology based on the soil moisture content calculation and reference evapotranspiration (ETo).

Chapter 6 reviews and focuses on several previous studies that have been done to study the effect of deficit irrigation on the yield of Sunflower. Special attention is given to deficit irrigation strategies in the field of water resources management.

Chapter 7 provides details about the field experiment methodologies of this study, which included three different irrigation treatments during a dry season as winter season and a rainy season as summer season.

The results of this research in Chapter 8 consist of three sets of findings. The Chapter presents first the results of the winter experiments, which were conducted during 2011, 2012 and 2013. The second set presents the results for the winter and summer seasons of the experiments of the intra-row plant spacing with three irrigation intervals and the analysis. The Chapter ends with the comparison between summer and winter seasons.

Chapter 9 briefly explores the water productivity with a focus on different terms of water productivity. Economic water productivity and economic analysis of different irrigation treatments are addressed in this Chapter.

Chapter 10 presents the description of the AquaCrop model for simulating water productivity and seed yield of Sunflower under different irrigation regimes and climatic conditions. Calibration and validation of the AquaCrop model were done for different irrigation regimes for all treatments. Moreover, the model performance was evaluated for the summer and winter seasons.

Finally in Chapter 11 the evaluation of this study and recommendations for future improvements of Sunflower cultivation under deficit irrigation conditions are presented.

2   Background and objectives

2.1   Country characteristics

Sudan lies in Northern Africa. It was ranked as the sixteenth largest country in the world and the third largest country in Africa (Figure 2.1). It has an area of 1,886,068 km² extending between 8° and 22° North latitudes and 22° to 38° East longitudes. It is bordered by seven countries such as Egypt in the North, Central African Republic, South Sudan and Uganda in the South, Libya and Chad in the West, and Eritrea and Ethiopia in the East. On the Northeast it is bordered by the Red Sea. The total population in Sudan was last recorded at 36.1 million people in 2015 with a growth rate of 1.72% (Central Intelligence. Agency, 2015).

Figure 2.1. Map of Sudan (www.wikipedia.org/wiki/Sudan)

In general Sudan has a tropical climate, from January to March the country is under dry conditions. The mean maximum temperature ranges from 30 to 40 °C and the mean minimum temperature from 10 to 25 °C during the winter and summer season. The average annual rainfall in the northern half of Sudan varies from 200 mm to 25 mm near the border with Egypt. The rainy season is limited to three months with the rest of the year almost dry in the centre of the country. The average annual rainfall which occurs between mid-June and September is about 254 mm. In the south of the country, the months of rainfall are concentrated between July and October. The average rainfall in this area ranges between 300 and 500 mm.

The topography varies from flat plains in the central part to low laying plateaus in some areas in the West and East. There are some hills in large masses such as the Nuba Mountains and the Ingessans Hills in the East, there are coastal plains of about 20 to 50 km wide covered with Red Sea Hills, which are rising to 1500 m+MSL (mean sea level) with 1000 m elevation of plateau to the West. In contrast there is a volcanic area in Western Sudan called Jabel Mara, which rises to more than 3000 m+MSL.

2.2   Water resources in Sudan

The Nile River, which flows from South to North towards Egypt, is the main source of irrigation water in Sudan. The long-term average annual flow of the Nile is 84 BCM (billion cubic metres) (Omer, 2008). According to the 1959 Nile Waters Agreement with Egypt, Sudan's share of water is about 18.5 BCM as recorded at Aswan in Southern Egypt. Presently, 94% of the abstracted water in Sudan goes to agriculture, 5% to human and animal consumption and 1% for industrial and other uses. Nile River has two main tributaries, Blue Nile and White Nile, which have to a varying degree influence on the agricultural production system (Figure 2.2).

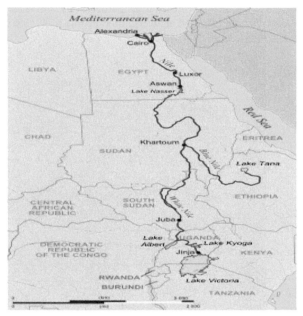

Figure 2.2. Map of the Nile River and its tributaries in Sudan
(Source: Wikipedia.org, River-Nile map.svg)

The Blue Nile has a total length of 1,450 km of which 800 km are inside Ethiopia. It flows from Lake Tana in the South to West across Ethiopia and then into Sudan from the Northwest side. After the flow passes El Roseires, it is joined by the Dinder River at Dinder Town. North of WadMedani El Rahad River joins the Blue Nile. After joining with the White Nile in Khartoum it flows as the main Nile. In Atbara Town the Atbara River joins the main Nile. The maximum discharge of the Blue Nile is in the rainy season from June to October and represents about two thirds of the water supply of the Nile River water. The Blue Nile and Atbara River were responsible for the annual Nile floods, which contributed to the soil fertility of the Nile Valley in Egypt.

The largest water user is agriculture with 36 km^3 from surface water resources. Groundwater is used only for domestic water use in limited areas. The irrigation potential is estimated at about 2.78 million ha based on soil and water resources criteria. Surface water provides water for 96% of the total irrigated area, and the remaining 4% is irrigated from groundwater (small tube wells). The irrigated area where pumps are used to lift water was 347,000 ha in 2000. The large-scale irrigation schemes under the government sector are managed by what is known as Agricultural Corporations, while small schemes are owned by farmers. The Gezira Scheme had many problems such as financial, technical and institutional problems, which resulted in a considerable fall in the crop production of the scheme and a corresponding drop in farm incomes in the late 1990s. The cropping intensity dropped from 80% in 1991/1992 to 40% in 1998/1999. Due to siltation and water mismanagement about 126,000 ha were taken out of production, resulting in a reduction of water availability. Because of water mismanagement, water supply is about 12% below crop water requirements at crucial stages in the growth cycle, while at the same time, as much as 20% of the water delivered is not used by the crops.

2.3   Crop production in Sudan

Population will increase (2.6% growth rate) and the living standards will improve. Therefore, the food demand will increase during the next decades. Most of this increase will be met by the products of irrigated agriculture. At the same time, water supply for the irrigated area will be reduced due to water scarcity and environmental changes. Water productivity is expected to increase through gains in crops yield and reduction in water applied. An increasing competition in water use raises the concept of better use and management of water resources, so that the need of all stakeholders can be met properly. Agriculture is being considered the major user of water.

In Sudan the cost of Sunflower seed is high, which provides an incentive to release new local varieties and to reduce planting rates and water application in specific crop stages to encourage farmers to produce more seed and enhance their income. Production of Sunflower has greatly increased in Sudan in the irrigated area of the central clay plain due to decline in the amount of rainfall in rainfed areas.

In Sudan the annual cereal harvest is estimated at 2.8 million tons. Due to poor rains the harvested areas were reduced in the rainfed sector. However, cereal production was less than half last years, which included Sorghum (2.1 million tons), Millet about 0.36 million tons and a low Wheat harvest of 0.32 million tons in March-April 2010. Oilseed production has decreased due to a reduction in Sesame production from 0.36 to 0.193 million tons, drop in Sunflower seeds from 0.122 to 0.09 million tons (FAO, 2011). Crop production in the rainfed areas is characterized by high annual fluctuation resulting from rainfall variation, while in the irrigated areas production and productivity levels are reasonably stable.

2.4   Increase in total crop production

To improve productivity there are many constraints. One of these is the limited knowledge of irrigated crop production among farmers. In Gezira Scheme irrigation efficiencies and system application were below the norm and irrigation scheduling did not take crop stages into account (personal communication). In addition to poor management, other main constraints to crop production are inadequate water supply and weed management. Regarding this, it is expected that farmers could benefit from the water users associations or block groups to face constraints in irrigation management.

Although crop productivity is affected by many factors, water is the major one. Potential water productivity is the ratio of crop produced under given water supply to yield that could be obtained if the same quantity of water was delivered under a supply system, which exactly matches the crop demand (García-Vila, 2012). Management of productivity can be a good parameter for describing how water supply can be adequate for satisfying the crop demands and how water can be used effectively.

The Gezira Scheme represents 47% of the total irrigated area and 10% of the total area under crop production in Sudan. The major problems facing the Gezira farmers include poor operation and management, shortage of equipment, shortage of skilled (manpower) labourers, commodity price controls, late delivery of inputs and lack of technologies for cultivation. There are also problems related to head - tail effects, where yields of crops are lower at the tail of the system (Table 2.1). FAO (2011) reported that the factors responsible for low yields in Gezira scheme include shortage in water supply due to poor and lack of maintenance of the water distribution system. Another reason affecting the crop yield is the head to tail effect which may cause inequity in water distribution.

Table 2.1. Yield effects on crops due to distance from the water source

Crop	Gezira Main	
	Head	Tail
Wheat (t/ha)	1.36	0.93
Groundnut (t/ha)	0.95	0.74
Sorghum (t/ha)	1.21	0.83

(Source: FAO, 2008)

2.4.1 Improving water use in crop production

There is increased competition for water between agriculture and other users, such as in the regions where rainfall is less than the crop water demand. Some cropping areas can experience regular droughts at one time in the growing season, whereas others have droughts at different growth stages (Morison *et al.*, 2008). In some areas yield depends on the water stored in the soil profile or rainfall only, while in other areas plant yield is a result of water available from soil, rainfall and irrigation. These variations of water mean that there are different targets to improve crop yield under water scarcity. Much effort has been made to reduce water use by crops and produce 'more crop per drop'. The improvements in both agronomic and physiological understanding have led to increased water productivity for some crops. To reduce agricultural water use and make water resources more sustainable, more combination efforts between agronomic, physiological, biotechnological and engineering approaches are needed. However, to develop crops that require less water to produce sufficient crop production research studies need to involve different disciplines. In contrast, to determine crop growth response to water availability and to reduce the amount of water used per unit yield research needs to be focussed on this topic.

2.5   Land use and land cover

The estimated arable land ranged from about 36 million ha in the mid 1960s to 84 million ha as published by the Ministry of Agriculture and Natural Resources in 1974. However, about 21.3 million ha is actually under cultivation, including about 10 - 11% fallow land. However, Sudan has about 120 to 150 million ha of which 76 - 86 million ha suitable for cultivation and grazing was estimated by the United Nations Development Programme (FAO, 2011). Table 2.2 shows the ecological zones in Sudan, which is divided into six zones. Dry and semi dry zones are the most important zones, which cover most of the agricultural areas in the country.

2.6   Problem description

Gezira Irrigation Scheme is fed by gravity from Sennar Dam on the Blue Nile at about 110 km south of the city of WadMedani. Due to the ageing of the system, siltation and weed growth, poor water management and the increased demand for intensification and diversification, which was introduced around 1975, water delivery at the farm level has become unreliable. Consequently the farmers tend to use water during the night for private crops. By so doing, they interrupt the water flow through the system. Further the behaviour of the farmers to practice unattended watering is believed to waste water. As a result, the agricultural sector suffers from major shortages of water supply at times, and inundation and waterlogging occasionally when rainfall is in excess and the drainage system fails to operate. Productivity of major food crops, namely Sorghum and Wheat has been very low at a maximum of 1 to 2 t/ha. Yield reductions are even more severe among other important food and cash crops (Cotton, Groundnut and Sunflower) as these have longer flowering periods and hence require a higher supply of water, and are more sensitive to drought spells.

In the past three to four decades, Sudan has witnessed increasingly varied and erratic rainfall. Because of this irrigated agriculture under water deficit irrigation has become one of the major production systems for ensuring food self sufficiency and food security. Accordingly different national irrigation and agricultural institutions, including the Agricultural Research Corporation, have been and are engaged in maximizing productivity and profitability of the major crops under water scarcity conditions.

Sunflower was proposed for Gezira Scheme as a new crop. This research is one of the initiatives. Its central aim is to investigate through field experiments the yield response of Sunflower (two varieties) to different water applications and to recommend the best irrigation and farming practices that will result in higher yield per given quantity of water and initial investment. Although Sunflower is increasingly viewed as the most important oil crop its yield has remained low, less than 2.0 t/ha.

2.7   Hypothesis

The cropped area and crop rotation systems in irrigated agriculture change as well as the crop water requirements for different crops. This results in challenges to increase water productivity. Current water allocations in irrigated agricultural areas, especially in the Gezira Scheme throughout the growing seasons, are not fair because of some problems such as poor performance, mismanagement and shortage in water supply during the season. Due to these problems the agricultural production, particularly agricultural water productivity in Gezira Scheme is low and can be significantly improved.

In order to improve the water productivity, a comprehensive water use analysis is required to understand and quantify different water balance components and their short-term as well as long-term impact on agricultural production of the system. Various approaches such as field measurements and simulation modelling are investigated to quantify the water productivity values of irrigation systems.

Therefore the hypothesis is that the opportunity for increasing water productivity under water stress conditions is contingent on the determination and accurate implementation of the irrigation needed at each growth stage on time. In addition better crop management, irrigation scheduling, change in irrigation technology and servicing of irrigation equipment can increase water productivity. This will be of great help to further increase water productivity under deficit irrigation strategies.

Water productivity of Sunflower under different irrigation regimes on Gezira clay soil

Table 2.2. Ecological zones of Sudan

Zone	Part of Sudan (%)	Mean annual rainfall (mm)	Wet season	Dry season	Main land use types
1- Dry	28.9	<75	July - September	October - June	• irrigated agriculture • grazing along seasonal water course
2- Semi-dry	19.6	75 - 300	July - September	November - June	• irrigated agriculture • dry land farming in conjunction with water harvesting • pastoral
3- Low rainfall savanna	27.6	300 - 800	May - September	November - April	• irrigated agriculture • rainfed traditional cultivation • mechanized farming • pastoral • forestry
4- High rainfall savanna	13.8	800 - 1500	April - October	December - February	• traditional cultivation • pastoral • wildlife
5- Flood region	9.8	600 - 1000	May - October	December - April	• rainfed traditional cultivation • pastoral • wildlife
6- Mountain vegetation	0.3	300 - 1000	Variable	Variable	• traditional cultivation • pastoral • forestry - horticulture

(*Source: Adapted from Zaroug, 2006*)

2.8   Research questions

Deficit irrigation strategies were applied at different crop stages to identify which is the most crop stage sensitive to water deficit and to what extent the crop would be tolerant. However, the following scientific research questions remained and were addressed in the study:

- what is the actual water use and maximum water requirement for Sunflower?
- what is the optimum water requirement for Sunflower to attain maximum yield?
- what are irrigation regimes that can result in maximum yield if they are applied?
- which irrigation regimes will significantly affect yield components?
- what is the effect of water stress on oil seed content?
- what are the best agronomic practices to produce optimum yield?

2.9   Research objectives

The main objective of this study was to determine the optimum crop water requirement and crop coefficients of Sunflower under different irrigation regimes under Gezira Clay Soil conditions. This study also covered the following specific objectives:

- to investigate the impact of water stress on yield and yield components as well as on oil yield content under different irrigation regimes;
- to determine the water productivity under different irrigation conditions;
- to assess yield response to different irrigation levels by using the AquaCrop model;
- to formulate recommendations on future cultivation practices of Sunflower in Gezira Scheme.

3 Agriculture in Sudan

Agriculture contributes about 40% of the GDP for the Sudanese people, which provide employment for about 80% of the labour force. Within the agricultural sector, crop production accounts about 53% of the agricultural output, 38% of livestock and 9% of forestry and fisheries (IFAD, 2002). The agricultural sector in Sudan is divided into three sub-sectors.

- traditional rainfed agriculture;
- mechanized rainfed agriculture;
- irrigated agriculture.

3.1 Agricultural practices

In Sudan agricultural practices or farming systems are a function of ecological zones and socio-economic conditions. There are three main farming systems for different crop production as shown in Table 3.1.

Table 3.1. Agricultural sector and main crop areas

Subsector	Area (million ha)		Main crop
	1995/1998	2001/2002	
Irrigated farming	1.63	3.78	Sorghum, Wheat, Cotton, Sunflower, Groundnuts, vegetables, fruit trees, Alfalfa, Forage Sorghum
Traditional rainfed	8.21	9.12	Sorghum, Millet, Sesame, Groundnuts, Watermelon, Roselle, Cowpea
Mechanized farming	7.93	12.80	Sorghum, Sesame, Cotton, Guar, Sunflower

(Source: Ministry of Agriculture and Forestry, 2008)

3.1.1 Rainfed agriculture

Traditional rainfed agriculture

Traditional rainfed farming occupies an area of 52.4 million ha, which is operated by the rural people. Most of this area is in Kordfan, Darfur, White Nile and Blue Nile States. It mostly consists of 10 - 50 ha small family units, farming for both income and subsistence. Farmers in this sector use good farming practices with wider use of crop rotation. These farms have a high production compared to the mechanized sector. For example 95% of the pearl millet, 38% of the Sorghum, 67% of the Groundnut and 38% of the Sesame is grown here (FAO, 2011).

Mechanized rainfed agriculture

Semi mechanized farming is made up of individual large-scale farmers and companies. In this sector, low-cost, soil mining, low-input agriculture has been adopted for over 50 years, with rainfall in scattered locations leading to low yields of crops. Rainfed commercial semi mechanized systems, with a total area of about 11.8 million ha, is localized in El Gedaref, Blue Nile, White Nile, Sennar, Southern Darfur and Southern

Kordfan. In this sector the production is low due to low yields over the past ten years, especially of Sorghum.

3.1.2 Irrigated agriculture

Irrigated agriculture is the major consumer of fresh water, as around 70% of all freshwater withdrawals are used for food production (Calzadilla *et al.*, 2010). Over 40% of global food is produced from irrigated agriculture. In the future, however, the necessary growth in food production has to be achieved with less water. So an increase of agricultural output per unit of water i.e. more crop per drop will be required (Kijne *et al.*, 2003b). Within this context deficit irrigation has widely been investigated as a valuable production strategy in dry regions (English and Raja, 1996; Pereira, 2002; Zhang, 2003).

In Sudan the irrigated agriculture area covers about 2 million ha, and contributes about 64% to the Gross Domestic Product (GDP). The source of water for this area is mainly the River Nile either by gravity irrigation from Dams or by means of pumping or by flood irrigation from the Gash and Baraka rivers. The main crops in the irrigated area are Sorghum, Groundnuts, Sugar Cane, Cotton, Wheat and vegetables. Within the Nile River Basin there are 1.26 and 1.68 million ha of irrigated land, mainly in North Khartoum, Gezira, Sennar, Blue Nile and White Nile State. Table 3.2 shows the average yield for irrigated crops in Sudan (FAO, 2008). The average yield of the cereal crops in Gezira Scheme is low 1.5 and 1.6 t/ha for Sorghum and Wheat respectively, while the yield from research trials is 3.8 - 5.7 t/ha for Sorghum and 3.3 - 4.1 t/ha for Wheat.

Table 3.2. Average yield of irrigated crops

Crop	Average yield (t/ha)	Research trials (t/ha)
Sorghum	1.5	3.8 - 5.7
Wheat	1.6	3.3 - 4.1

(Source: FAO, 2008)

3.1.3 Flood/spate irrigation agriculture

Spate irrigation in Sudan has been practiced in several areas for many years. In Eastern Sudan, the Gash Scheme in Kassala State is an example of a large scheme of spate irrigation. The total area is 280,000 ha, in which agriculture covers about 180,000 ha and 100,000 ha remain to be irrigated by a canal network (Hamid *et al.*, 2009). Sorghum is the main cereal crop, it contributes 40% and about 60% is cash crops. In the Gash Delta the river is the major source of groundwater, where the water depth varies between 5 - 18 m. It rises from July to September during the flooding period and it reduces in May or June.

3.2   Water management for agriculture

Irrigated agriculture is the bulk of the water demand in Sudan, which usually is affected by water shortage. In order to sustain its needs, we need to focus on the efficient use of all water sources (surface water, ground water and rainfall), and on the strategies of water allocation that maximize the economic return to limited water availability and hence improve the water productivity. In this endeavour, there is a need to focus on water management for agriculture and policy issues. Increasing the efficiency of water use and enhancing agricultural water productivity at different agricultural sectors is becoming a priority in Sudan. Irrigation management is important in Gezira Scheme, where the different types of crops are cultivated in a large part of the scheme. The main challenge facing water management in agriculture is to improve the efficiency of water use. Water management for crop production in large-scale schemes is critical.

3.3   Agricultural policies and strategies

The agricultural sector in Sudan has suffered in the last 20 years from policy changes and low productivity. Mainly in Gezira Scheme, due to instability of agricultural policies from the Gezira Land Act of 1927 up to the Gezira Act of 2005, productivity for main crops decreased with time (Raes *et al.*, 2009). The policies in Gezira Scheme were changed over time from direct responsibility of the government towards more WUAs participation (Personal communication, 2013). The major problem of irrigation management in Sudan is low productivity. Therefore, more attention to these changes in productivity would have to be taken into consideration.

The present policy system adopted several new policies to save the Sudanese economy through agricultural development and production. A lot of efforts and investments were focused on and towards rural development, which is the objective of economic salvation. These policies are geared towards multiplying agricultural production to satisfy the local needs. One of these agricultural policies to achieve food security is by increasing lands for growing Wheat and concentration on the cultivation of Sorghum in the irrigated areas to boost production. Moreover, there was an expansion of rainfed land. Also, there was a focus on small-scale production as standard practice for agricultural development. Efforts were increased to diversify and add new crops such as fodder, Sunflower and Soybean.

3.4   Sunflower (Helianthus annuus L) description

Sunflower (*Helianthus annuus* L) is the most important source of edible vegetable oil after Soybean, Rapeseed and Peanut, with a worldwide annual seed production of 33.3 million tons destined almost exclusively to oil extraction, providing 8.5% of the total world volume (FAO, 2008). However, Sunflower production and yield stability is limited by drought stress, especially in arid and semi-arid areas. Water stress on Sunflower has been reported to reduce plant height, root length, dry matter and grain numbers per plant (Nezami *et al.*, 2008; Nazariyan *et al.*, 2009). Hysun 33 is a crop variety having medium tolerance to water stress (Ahmad *et al.*, 2009). It has been found that both quantity and distribution of water has a significant impact on seed and oil yield of Sunflower (Iqbal *et al.,* 2005). The oil content in Sunflower seed ranges between 35 to 50% and consists for about 90% of unsaturated fatty acids, placing it as one of the best healthy oils for popular consumption (Shayesteh, 2013). Sunflower is a temperate area crop, but it can perform well under a variety of climatic and soil conditions. This crop is adapted to relatively low rainfall areas receiving winter and spring rainfall with a low humidity during flowering and maturation. Sunflower is an important oilseed crop because, it has a wide adaptability with different climatic conditions, suitability to mechanization, low labour needs and high protein and oil contents (Canavar *et al.*, 2010). It has an important role in crop rotation systems. Sunflower seeds have a high medical use, which can be applied as an addition to therapy of colon cancer, high blood pressure and migraine headaches. Sunflowers has a short growing season and thus lower irrigation needs, which has helped plant breeders to have more interest in this crop in different regions. It is thought to better withstand short periods of crop water deficit than Maize and Soybean. However, the timing of critical stages for water needs is offset from those of these two crops.

3.5   Sunflower growth stages

Sunflower has several growth stages such as emergence, vegetative, reproductive, flowering, seed formation and physiological maturity. Water stress in each stage results in reduction in seed yield and oil content. The treatments, which were conducted to study the effect of water stress on growth stages showed that Sunflower was significantly affected by water stress that occurred in the sensitive flowering, seed formation and ripening stages. The time required for growth development and time between various stages of development depends on the genetic background of the plant and growing season environment. Figure 3.1 shows Sunflower growth stages, which describes five stages. The initial stage includes two phases: germination and seedling establishment. This stage is sensitive to weeds and needs good control. After 20 to 30 days after sowing (DAS) crop development starts till 45 days. During this stage the flower bud stage starts and at the end of this stage (11 days) it is the time of initiation of the flowering stage. This stage is more sensitive to water stress and is considered as a critical stage. The late stage is the stage before physiological maturity, which is the seed filling stage and it is sensitive to water stress. The physiological maturity stage starts when the colours of the leaves and heads change to brown and when they become dry.

3.6   Sunflower seed production

Globally, Sunflower production is around 35.6 million tons with 25.1 million ha of production area (FAO, 2008). The most important Sunflower-producing countries in the world are Russia, Ukraine and Argentina, while Russia is the largest producer of Sunflower seed in the world. Although in India the harvested area of Sunflower is larger than in the other countries, the yields are low compared to the other countries. In Sudan Sunflower seed production was 92,000 tons and harvested area about 99,960 ha in 2012 (FAO, 2008). In Gezira and Managil schemes the Sunflower planting areas dropped from 21,000 ha in 1994/1995 to 12,600 ha in 2012/2013 and seed yield production decreased to this reason from 14,000 tons to 2000 tons (Tables 3.3 and 3.4) (Ministry of Agriculture and Forestry, 2008).

Table 3.3. Sunflower production in Sudan (tons seeds) from 1995 to 2006

Seeds (ton)	Year	Change compared to previous period (%)
14,000	2006	117
12,000	2005	171
7,000	2004	38.9
18,000	2003	94.7
19,000	2002	475
4,000	2000	16
25,000	1995	-

(Source: Statinfo.biz/Sudan/agriculture, International economic statistics.2011)

Figure 3.1. Sunflower growth stages at the experimental plots

Table 3.4. Sunflower seed production in Gezira and Managhil schemes

Year	Planted area (1000 ha)	Harvested area (1000 ha)	Production (1000 ton)	Yield (kg/ha)
1994 - 1995	21.0	21.0	14	688
1995 - 1996	0.4	0.4	-	3570
1996 - 1997	6.0	6.0	2	793
1997 - 1998	0.4	0.4	Nr	666
2005 - 2006	0.4	0.4	Nr	1600
2006 - 2007	2.5	2.5	3	3810
2012 - 2013	1.3	1.3	2	1590

(Source: Ministry of Agriculture, 2014). Nr: data not recorded.

4 Irrigation schemes in Sudan

4.1 Introduction

FAO estimated that the irrigated area in the selected 93 developing countries will only rise by 23% over the 1998 - 2030 periods. The effective harvested irrigated area considering the increase in cropping intensity is expected to increase by 34% (Playán, 2006). This means that more area needs to be brought under irrigation to produce more crops with the present available water resources.

Sudan has an agricultural potential of 105 million ha, of which only 16.7 million ha are cultivated and about 1.9 million ha out of an irrigation potential of 2.8 million ha is under irrigation now. Therefore, there is wide room for further developments, especially in the irrigation sector. However, there are three major constraints to irrigation development in Sudan:

- ineffective process of annual maintenance and operation of canals, especially for the removal of silt due to lack of funds;
- a steady increase in development costs, which is affected by international trade;
- lack of farmer participation in the planning and operation of the schemes or other related services.

The development plans for the irrigation sector include the rehabilitation of the existing irrigation schemes, a shift of emphasis towards the development of small-scale irrigation schemes, phased development and vertical expansion. Agricultural sectoral policies for the irrigation sector include the following:

- to extend the market economy to all crops to allow the farmer to choose a suitable crop mix;
- to provide support for the agricultural research institutions so as to explore suitable technologies for improving crop production and productivity. In addition, to provide support to agricultural extension and services;
- to encourage exports through providing facilities to agricultural companies;
- establishment of specialized crop committees for main crops;
- farmers to participate in formulation of agricultural policy.

The development efforts in water resources have been implemented through a participatory approach. Certain parts of irrigation systems were maintained by government and a lower part by farmers. The distribution of water from the canals is the responsibility of the farmers, which caused some problems in the systems as follows:

- water supply to farmers is increasingly unreliable and inequitable;
- deferred maintenance is resulting in poor performance in the irrigated sector;
- degradation of land is taking place due to excessive irrigation by some farmers;
- the cropping pattern is not decided on the basis of total available water.

Generally, the water supply for the irrigated schemes is provided from reservoirs and/or pumps and extensive networks of canals covering the whole schemes as well as drainage networks. The overall objective of water management policies is to improve the water productivity in agriculture, which includes efficient control of water in the irrigation systems, efficient and economical maintenance of the irrigation canals and structures, and technical capacities to operate the systems. Supplementary irrigation could increase the very low or zero productivity of crops and fodder in the marginal rainfall areas. The efficient use of groundwater could help to optimize the water resource productivity.

4.2   Surface irrigation

Surface irrigation is the traditional irrigation method in Sudan and mainly in all the irrigation schemes which based on providing sufficient water to avoid water deficit at all times, so as to obtain maximum yields.

4.2.1 Supplemental irrigation (SI)

The rainfed areas play an important role in the production of food in many countries, 80% of the land produced 60% of the total production used for cropping throughout the world. Due to variability in amount and distribution, however, the rainfall in most areas does not meet the seasonal crop water requirements. It was found that yields and water productivity (WP) was enhanced by the use of rainfall with limited irrigation water. Many studies on supplemental irrigation (SI) showed an increase in crop yield in response to application of a small amount of water as supplemental irrigation. Therefore SI is important for water productivity. Water productivity of both irrigation water and rain water is improved when SI is applied. In rainfed areas, strategies to optimize deficit irrigation need information about the amounts and distribution of rainfall, in addition the sensitive crop growth stages to moisture stress. Oweis (2006) reported that in rainfed areas, supplemental irrigation optimization is based on two basic aspects:

- water added to rainfed crops encourages producing some yield;
- SI would have to be applied to rainfed crops when rainfall fails to meet the essential water moisture to improve yield production.

4.3   Irrigation scheduling

Irrigation scheduling is the water management strategy of when and how much water to apply to an irrigated crop to maximize total yield. Commonly irrigation scheduling is defined as determining the time of irrigation and the amount of water to be applied. The maximization of yield requires a high level of water productivity. This requires the accurate measurement of the volume of water applied or of the depth of application. Under the water stresses there are reductions in yield and decrease in yield returns. To gain benefit from irrigation scheduling, there must be an efficient irrigation system. Irrigation scheduling is one of the important management activities that are vital to the effective and efficient utilization of water. Water management strategies based on irrigation scheduling are intended to reduce the amount of water applied and to minimize yield reduction due to water stress. It will ensure that water is applied to the crop when needed and in the amount needed.

Because of the need to optimize yield under irrigated agriculture, irrigation investments are top priority. Hence, it became a matter of serious concern in recent years. Irrigation practices will be used at most exclusively to supplement the amount of soil moisture stored in the root zone to satisfy the crop throughout the growing season. As a result, even today, farmers still tend to over-irrigate to ensure a satisfactory amount of water stored in the soil profile. However, farmers may deal with irrigation scheduling and integrate other agronomic practices, such as fertilization and pest control only if the irrigation delivery can be predicted.

Although irrigation scheduling is very important yet there are many reasons and constraints that may hinder farmers and engineers to use it, which include unreliable weather prediction and time required. The approaches needed to conduct irrigation scheduling are water use and weather prediction, soil water monitoring and crop water stress monitoring. Soil water monitoring is tedious, labour and equipment intensive. The crop water stress technique is a useful method to improve understanding of plant water stress. Irrigation scheduling based on weather prediction remains the most practical and common method. Determination of the reference evapotranspiration (ETo) is the most common method to estimate crop water use from the local climatic data and crop coefficient for different crop stages (Allen, 1998) Irrigation scheduling is the most important factor affecting irrigation efficiencies and crop yields. When irrigation water is short with high costs, irrigation scheduling is important when soil moisture conditions generally restrict water movement or development of the crops (Jensen *et al.*, 1970). For achieving high efficiencies, however irrigation scheduling practices must be improved.

Studies dealing with the effect of various agronomic practices on Sunflower yield and yield components have greatly increased in the central clay plains, such as intra-row plant spacing, row spacing and irrigation scheduling (Ahmed, 2010). A better understanding of these agricultural management matters could lead to improving the Sunflower management for good seed and oil production. Recently Sunflower has been grown as a commercial crop and oil seed crop in the dry and semi-dry regions of Sudan. In the areas where precipitation is limited, irrigation scheduling is important to increase production.

Where rainfall is not sufficient and irrigation water supply is restricted, irrigation to attain optimum production would have to be based on avoiding water deficit during the periods of peak water use from flowering to early yield formation. An understating principle of irrigation scheduling helps farmers to develop their own strategies for their specific regions and condition.

4.4   Water balance

The water balance for the root zone concerns the amount of water entering down and moving up of the effective root zone of a crop. It is based on estimating the soil moisture content in the root zone viewed as a system. Water added to the root zone by irrigation and rainfall can be lost before entering the root zone by runoff or percolated below the root zone (Figure 4.1). Water is also lost from the root zone through the evapotranspiration from soil and plants. The total soil moisture content in the root zone in any day can be calculated by using the following equation:

$$TW_T = TW_{T-1} + I + P \pm ETc \pm D \pm R + F_{net} \qquad (4.1)$$

Where:
TW_T = total soil water content in the root zone on day T
TW_{T-1} = total soil water content in the root zone on the previous day (T-1)
I   = irrigation water applied (mm)
P   = Precipitation or rainfall (mm)
ETc   = Evapotranspiration (mm)
D   = drainage or deep percolation (mm)
R   = runoff (mm)
F_{net} = any change in total water in the root zone from underground water movement

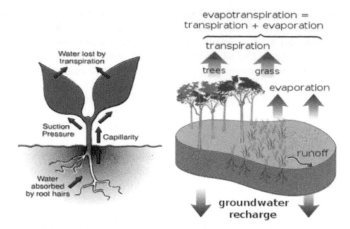

Figure 4.1. Transpiration and evapotranspiration cycle (Source: www. ENORASIS.eu)

Infiltration is the process by which water on the ground surface enters the soil. It is the measure of the rate at which soil is able to absorb rainfall or irrigation. It is measured in millimetres per time unit. At saturation conditions the rate will decrease until the rate becomes constant. If the precipitation rate exceeds the infiltration capacity, runoff will occur. In Gezira clay soil the infiltration rate starts high immediately after irrigation, decreases continuously with time and becomes constant after 2 or 3 days after the soil has reached the field capacity (FC).

4.5　Furrow irrigation

Furrow irrigation in Sudan is concentrated in the central clay plains in the Gezira area, where the Gezira Scheme is the largest scheme using gravity irrigation with water abstracted at the Sennar Dam.

4.6　Main irrigation schemes

In Sudan irrigated farming systems occupy about 1.86 million ha. Four schemes, Gezira, Rahad, New Halfa and El Suki, are the largest national schemes of the country. They cover more than half the irrigated area and consume around 60% of the Sudanese annual abstraction (between 13 and 16 Mm^3), totalling 1.2 million ha and irrigated by gravity supply systems (Table 4.1). The irrigation schemes are managed by the public sector except Kenana Sugar Scheme. There were more than thousand tenant farmers working in these schemes who cultivated all the Cotton of the country, most of its Wheat, 35% of the Groundnuts and 10% of the Sorghum (FAO, 2011).

4.6.1　New Halfa Scheme

New Halfa is one of the oldest schemes in Sudan. The total area is about 164,000 ha, irrigation by furrows is the common irrigation method in this scheme (Zwart and Bastiaanssen, 2004). It was considered as one of the largest population resettlement projects in the country. The Khasm El Girba Reservoir feeds the water to the scheme.

4.6.2   Rahad Scheme

The Rahad Agriculture Scheme was constructed during the mid 1960s. In 1981 the scheme was fully operated. Previous studies indicated that the scheme area is suitable for different cropping patterns. The total area is 126,000 ha and extends from South to North with 10 cm per km gentle slope. The soils in the scheme are heavy clay. The scheme lies in semi-dry zone. The rainy season extends from July to October with an average annual rainfall of 300 - 400 mm. Table 4.2 demonstrates that in the Rahad Scheme there are major gaps between the maximum and the potential yields and between minimum and maximum actual yields for the four crops.

Table 4.1. Government irrigation schemes in Sudan

Scheme	Equipped area (ha)
Gezira and Managil	870,750
White Nile pump schemes	192,375
New Halfa	152,280
Rahad	121,500
Blue Nile pump schemes	112,590
Gash Delta (spate irrigation)	101,250
Northern pump schemes	41,715
Suki	35,235
Tokar Delta (spate irrigation)	30,780
Guneid Sugar	15,795
Assalaya Sugar	14,175
Sennar Sugar	12,960
Khashm El Girba	18,225
Other areas	143,370
Total	1,863,000

(Source: FAO, 2011)

Table 4.2. Potential and actual crop yield in the Rahad Scheme, Sudan

Crop	Potential yield (t/ha)	Minimum actual yield (t/ha)	Potential (%)	Maximum actual yield (t/ha)	Potential (%)
Groundnut	0.76	0.16	21	0.34	45
Sorghum	0.40	0.12	30	0.28	71
Wheat	0.60	0.10	17	0.38	63
Cotton	3.6	1.2	32	2.9	80

(Source: FAO AQUASTAT, 2012)

4.7   Gezira Irrigation Scheme

4.7.1   Description of the scheme area

The location of Gezira Scheme between the Blue Nile and White Nile rivers is shown in Figure 4.2. The scheme area covers more than 300 km from South to North and 150 km from East to West, which constitutes more than half of the total irrigated area of 1.8 million ha (Elias *et al.*, 2001). The initial area was designed to serve about 126,000 ha (300,000 feddan) in Tayba Village (north-east of WadMedani). An extension to the scheme was executed in the late 1920s and early 1930s. In the early 1950s another extension was implemented to increase its size to 420,000 ha (one million feddan). Moreover, in 1957 a start was made with the Managil extension of about 336,000 ha (800,000 Feddan) that was completed in the mid 1960s (Sun *et al.*, 2006). It was designed to be irrigated by gravity from Sennar Dam on the Blue Nile, about 110 km south of WadMedani. Gezira Scheme has 47% of the total area under irrigated agriculture and produces 50% of domestic crop output (Guvele and Featherstone, 2001). Gezira Scheme plays an important role in the Sudanese economy and contributes about 35% of total GDP.

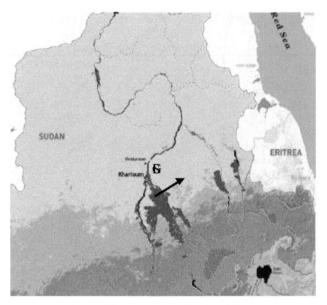

Figure 4.2. Location of Gezira Scheme

The water quality for irrigation from the Blue Nile is excellent. There is no effect of salinity on the scheme area except in the south of Khartoum. After the rainy season, sediment is the main problem in the minor canals, which need to be removed yearly. The soil has a high clay content with no deep drainage. The Sennar Dam was built to control the Blue Nile flow.

4.7.2   General climatic conditions

The Gezira Scheme area lies within the dry zone, which is characterized by low average annual rainfall of about 200 to 300 mm and clear fluctuation in the distribution and intensity of rainfall from year to year. Maximum temperature ranges from 34 °C in January to 41 - 42 °C in April and May, while the minimum temperature ranges between 14 and 25 °C in January and June respectively (Figure 4.3). The relative humidity is low most of the year, 50% is the mean relative humidity for the months between October and June (Table 4.3). The wind speed is also low; the range is 2 - 3 m/s at 2 m height. Evaporation is high during most of the year due to the high solar radiation; the highest evaporation is 8 mm/day in April-May, while the lowest is about 6 mm/day in December-January (Adam, 2014).

4.7.3 Design of Gezira Scheme

The Irrigation system consists of two main canals running from Sennar Dam south of Gezira State with total capacity of 354 m³/s. Total length of the network of branch canals and major canals is about 2,300 km. There are about 1500 minor canals with a total length of over 8,119 km. The major and minor canals are designed as distribution system canals (Plusquellec, 1990). Gezira Scheme system was designed to fit farmers and crop rotation. Fields with a gross area of 37.8 ha (90 feddan) are termed a number. In each number there are 18 farmer fields of 2.1 ha (5 feddan) each, which is referred to as Hawasha (Figure 4.4). It is allowed to grow one crop per year, followed by four rotation systems. Major and branch canals are considered as conveyance canals, while minor canals for permanent water flow from the major canals.

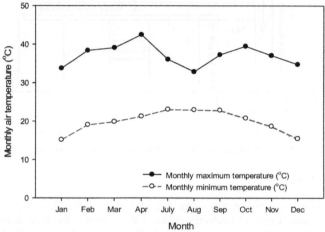

Figure 4.3. Monthly maximum and minimum temperature at Gezira Scheme (2012)

Table 4.3. Mean monthly climatic data of the Gezira Research Station (2012)

Month	Maximum temperature (°C)	Minimum temperature (°C)	Relative humidity (%)	Sunshine (h)	Wind speed (m/s)
January	33.8	15.2	31	10.5	2.3
February	38.4	19.1	28	10.3	2.5
March	39.1	19.9	24	9.8	2.5
April	42.5	21.3	23	10.5	1.8
July	36.1	23.1	62	6.5	4.2
August	32.9	23.0	67	5.8	3.3
September	37.3	22.9	57	10.1	1.6
October	39.5	20.8	40	10.1	1.6
November	37.1	18.7	40	10.5	1.6
December	34.8	15.5	30	10.7	1.8

(Source: WadMedani Meteorological Station, 2012)

The Gezira irrigation system is under control by two managements, irrigation authorities, Ministry of Agriculture and Irrigation (MAI), which has the responsibility to allocate water for irrigation and Sudan Gezira Board (SGB) that is responsible to deliver water to the farmer's fields as well as to determine crop water requirements. Water flow was controlled from major to minor canals by movable weirs.

Cropping pattern was changed in Gezira Scheme from four-course to five-course as crop rotation in 1991/1992 crop year after entered livestock and is as follows: Cotton - Sorghum, Groundnut - Wheat (as winter or fallow) (Table 4.4). Cotton grows as summer crop in designated area as decided by the Sudan Gezira Board (SGB).

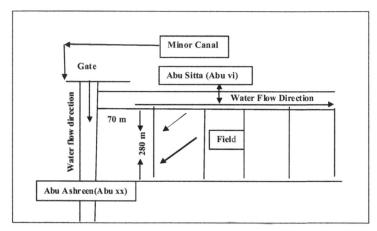

Figure 4.4. Canal system of Gezira Scheme for each field (Hawasha) 2.1 ha (5 feddan)

4.7.4   Gezira clay soil

The Gezira clay soil is heavy - cracking clay soil (Vertisol), it swells and shrinks upon wetting and drying respectively (Elias, 2001). Hence the bulk density changes with the change in moisture content and depth. There are no deep drainage losses in this soil. Mitchell (2010) described the Vertisols as dark, cracking, and montmorillonitic clays, known as black cotton soils in the Gezira.

The Vertisols are deep, dark coloured, low in organic matter very slowly permeable when wet and deeply cracked when dry. The clay content ranges between 50 and 60%, the top 60 cm is dark brown and the second 60 cm is grey, calcium carbonate and gypsum accumulation occur in the sub soil (Farbrother, 1996). Farbrother mentioned that, the norms of behaviour of Gezira clay soil (Vertisol) under irrigation was described as the eccentric 'odd man-out', when compared with the norms of behaviour of Vertisols under irrigation elsewhere in Africa, Asia and Australia. It is certainly so far as they have been identified (Farbrother, 1996). He stated that the main feature of soil resulting from irrigation moisture distribution following irrigation, the soil moisture increases to its greatest level in the top 20 cm and fells thereafter in the 60 - 80 cm zone. The Vertisols have cracks, which are most important phenomena of the clay soil behaviour. Cracks appear in the dry season in Gezira heavy clay soils.

Table 4.4. Cropping calendar in Gezira Scheme

Crop	Jan	Feb	Mar	April	May	June	July	Aug	Sep	Oct	Nov	Dec
Cotton	Irrigation			land preparation			planting			irrigation		
	Harvesting											
Wheat		Harvesting		Off season				land preparation			planting	irrigation
Groundnut						planting						
Sorghum		Off season		land preparation			planting	irrigation		harvesting		
										irrigation		

(*Source: Agricultural Research Corporation*)

4.7.5   Water application management in the field

The traditional method of field water application is basins (by angaya or by hod). The networks of cross-bunds of the angaya system prevent the removal of water which may stand for days. Poor irrigation management, excessively wet seasons, and poor irrigation and drainage lay out can contribute to waterlogging and therefore reduce yields on Gezira clay soils.

Plusquellec (1990) gives waterlogging a substantial importance altogether beyond its practical relevance to water management in the Gezira. Good irrigation management, which aims at providing the crop with enough water to avoid stress while at the same time avoiding over watering and waterlogging is difficult to achieve in the soil and climatic conditions of the Gezira Scheme. Huygen (1995) and Farbrother (1996) reported that the objective of irrigation management is to maximize net return, minimize irrigation costs, maximize yield and allow for optimal distribution of a limited water supply. The main challenge confronting water management in agriculture is to improve water productivity and its sustainable use.

Inequity is a common problem between users in many irrigation systems around the world. It is commonly known as the top-tail ends problem (García-Vila, 2012). Inequity problems at the minor canal level are always associated with water shortage, because of reduced water supplies and increased competition for the water.

4.7.6   Water users associations in Gezira Scheme

A water users association (WUA) is a group of water users, such as irrigators who pool their financial, technical, material and human resources for the operation and maintenance of a water system. Originally WUAs were irrigation groups that worked in traditional irrigation systems. They are fundamentally based on a participatory, bottom-up concept. In 2005 the concept of water users associations was established by the Sudanese government in Gezira Scheme (Gezira 2005 Act) as a new step to improve irrigation system operation and maintenance. Accordingly, the responsibilities of irrigation water management and crop pattern choices at the farm level have been transferred to the WUAs. Many studies conducted in the scheme showed that the performance of Gezira Scheme declined in recent years, especially with respect to water management. It is indicated that the average water productivity decreased from 0.25 to 0.16 kg/m^3 under WUAs conditions (Eltigani, 2014). In principle, the benefits of collective action and the settlement of local disputes over water would have to be largely contained within the WUAs. However, the evidences show that so far in Gezira Scheme they only share a few features, and, more importantly, do not share the major ones to which they have limited contributions such as:

- to recognize to the policies of how to deal with all kinds of issues;
- to decide on how much and where they are going to pay for the water, and how the water fees can be collected;
- to easier mention production problems towards government representatives;
- to better coordinate the purchase agricultural inputs in bulk at lower prices;
- to be a platform for conflict management between head and tail of canal farmers.

5 Crop coefficient (K_C) for irrigated Winter Sunflower (*Helianthus annuus L*)

5.1 Introduction

Yield and quality of the Sunflower are negatively affected by insufficient water supply and improper scheduling of irrigation. Available irrigation water has to be utilized in a manner that matches the water needs of the crop. The water requirement of crops varies substantially during the growing period, mainly due to variation in crop canopy and climatic conditions (Doorenbos and Pruitt, 1977). The knowledge of crop water requirements is an important practical consideration to improve water productivity in irrigated agriculture. There is considerable scope for improving crop water use of irrigated crops in Gezira Scheme, mainly the cash crops (Cotton, Groundnut and Sunflower), by proper irrigation scheduling, which is essentially governed by evapotranspiration (ETc). Accurate estimation of evapotranspiration is therefore an important factor in efficient water management.

The water required by a crop is most important because water has a direct effect on the yield of the crop. Anwar *et al*. (1995) stated that the yield and yield components of Sunflower are affected by the number of irrigations. As the number of irrigations increased the days to maturity, seed yield and plant height increased. As agriculture is the single largest water user in Sudan, improved irrigation management could result in large water saving. Determining crop water requirements is the first step in reducing water use, while maintaining profitable production.

Values of the crop coefficient for a number of crops grown under different soil and climatic conditions have been developed by Doorenbos and Pruitt (1977). These values are commonly used in places where local data are not available. They also emphasized the need to develop crop coefficients under given climatic conditions. The crop coefficient (Kc), which is the ratio of ETc to reference evapotranspiration (ETo), is often used. Determination of the crop coefficient under local climatic conditions is the base to improve planning and efficient irrigation management for many field crops. In Gezira, crop coefficients for Cotton, Sorghum, Groundnuts and Wheat have been determined by Farbrother (1973). Since there was no study conducted to determine the crop coefficient of Sunflower in Gezira Scheme, this study will be useful for managing, planning and for identifying the water demand in the growing seasons.

5.2 Study area

The experiments were conducted at the Gezira Research Station Farm (GRSF), which is located at latitude 14.4° N and longitude 33 5° E in WadMedani in Central Sudan, at an altitude of 405 m+MSL (mean sea level). All agronomic practices were conducted during the growing seasons of 2011/2012, 2012/2013 and 2013/2014. Sunflower, variety Hysun 33 was sown in mid November in the first two seasons and on the first of December in the third season. The seed rate was 3 seeds per hole at 30 cm inter-row spacing and 80 cm between the rows. The plants were then thinned to one plant per hole after two weeks. The experimental design adopted was a randomized complete block design. Fertilization was applied as Urea at the rate of 46% N, 86 kg N /ha in two split doses, one with the second irrigation and the other after flowering as recommended by the Agricultural Research Corporation. The plot size was 50 m² (10 × 5 m). After emergence the field was irrigated every week up to the flowering stage and then on the basis of water deficit at three irrigation levels. The irrigation treatments were seven by using the furrow method, including irrigation every week (W), 10 days (F1), 15 days (F2) and 20 days (F3) intervals at the beginning of the flowering stage till physiological maturity, and 10 days (S1), 15 days (S2) and 20 days (S3) during the seed filling stage to maturity replicated three times. Daily weather data were collected from the nearby meteorological observatory of Gezira Meteorological Station in WadMedani.

5.3   Calculation method

An accurate calculation of daily evapotranspiration could improve irrigation water management, which requires accurate irrigation scheduling. Irrigation that is correctly timed minimizes over-irrigation as well as under-irrigation. FAO paper 56 describes a widely used approach to estimate water requirements of agricultural crops under different climatic conditions (Allen, 1998). All the weather data used to calculate the evapotranspiration were collected from WadMedani Meteorological Station located near the Gezira Research Station Farm (GRSF).

5.4   Reference evapotranspiration (ETo)

Reference evapotranspiration (ETo) associated with the crop coefficient is one of the most widely known and employed ways of evaluating water consumption by irrigated crops. ETo is a climatic parameter which can be computed from weather data (Amir Kassam, 2001). From meteorological data net radiation, soil heat flux, air temperature, wind speed, saturated and actual vapour pressure are used to calculate reference evapotranspiration. Since 1990 FAO recommends the Penman-Monteith equation as the standard to estimate the reference evapotranspiration. There are many equations to estimate ETo but the Penman-Monteith equation is considered as the most precise one. The equation as recommended by Allen (1998) is written as follows:

$$ET_o = \frac{0.408\Delta(Rn - G) + \gamma \dfrac{900}{T + 273} u_2(e_s - e_a)}{\Delta + \gamma(1 + 0.34u_2)} \tag{5.1}$$

Where:
ETo   = reference evapotranspiration (mm/day)
Rn   = net radiation at the crop surface (MJ m^2/day)
G   = soil heat flux density (MJ m^2/day)
T   = mean daily air temperature at 2 m height (ºC)
u2   = wind speed at 2 m height (m/s), es the saturation vapour pressure (kPa)
ea   = actual vapour pressure (kPa)
es - ea = saturation vapour pressure deficit (kPa)
Δ   = slope vapour pressure curve (kPa/ºC)
γ = psychrometric constant (kPa/ºC)

Figure 5.1 shows the calculated ETo by ETCalc version 3.1 as recommended by FAO (2009). ETo was high in the first season 2012 compared to the second season (2013). Table 5.1 shows the monthly climatic data for the three growing seasons, which were obtained from the nearest meteorological station. From Table 5.1 the calculated evapotranspiration increased with an increase in wind speed. Whereas the reference evapotranspiration (ETo) increases with an increase in wind speed in hot dry weather conditions. The ETo demand is high in the hot dry weather due to the dryness of air and the amount of energy available as direct solar radiation and latent heat. Under arid condition when the weather is hot and dry a small variation in wind speed may cause a large variation in the evapotranspiration rate.

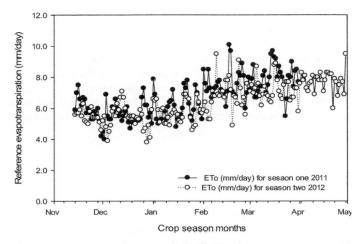

Figure 5.1. Reference evapotranspiration (ETo) for two growing winter seasons 2012 and 2013

Table 5.1. Mean monthly weather data of Gezira Research Station for the three winter growing periods (2011, 2012 and 2013)

Month	Maximum temperature (°C)	Minimum temperature (°C)	Relative humidity (%)	Wind speed (m/s)	Sunshine (hour)	ETo (mm/day)
2011/2012						
November	36.1	15.5	35	1.9	11.2	6.0
December	35.3	15.2	31	1.9	10.9	5.6
January	33.8	15.3	31	2.3	10.5	6.1
February	38.4	19.1	28	2.5	10.3	7.6
March	38.9	19.8	24	2.5	9.9	8.0
2012/2013						
November	37.2	18.6	40	1.7	10.3	5.6
December	34.8	15.9	31	2.8	10.7	5.4
January	34.8	17.3	33	2.8	10.2	5.8
February	37.0	20.1	27	2.3	10.2	6.8
March	40.5	19.3	23	1.8	10.4	7.2
2013/2014						
December	33.6	16.5	32	1.9	10.7	5.8
January	34.2	15.5	35	2.7	10.7	5.8
February	35.4	20.9	34	2.3	10.5	6.4
March	39.5	22.3	30	2.3	9.6	7.2
April	41.3	25.6	31	3.5	7.5	8.1

(Source: WadMedani Meteorological Station, 2014)

5.5   Evapotranspiration (ETc)

Water use by crops is a complex process, which includes losses by transpiration (T) from plants and evaporation (E) from the soil and from the plant's surface. It is difficult to measure evapotranspiration because it is difficult to separate transpiration from the plant and evaporation from soil. The actual evapotranspiration was calculated by the water balance equation as follows:

$$ETa = I + P \pm Dp \pm R - \Delta S \tag{5.2}$$

Where ETa is the actual evapotranspiration (mm), I is the irrigation water applied (mm), P is the precipitation (mm), Dp the deep percolation below the root zone (mm), R the surface runoff (mm) and ΔS the change in soil moisture storage (mm) to a profile depth of 100 cm. However, in the dry season P is zero, Dp can be neglected in Gezira clay soil, R (was assumed to be zero) (Adam, 2014). Then the water balance equation can be reduced to:

$$ETa = I + \Delta S \tag{5.3}$$

5.5.1 Soil moisture measurements

The classical method for determining soil moisture is the gravimetric method, which is accurate and simple where many locations can be sampled. The gravimetric technique is probably the most widely used for measuring soil moisture and is the standard method for the calibration of all other soil moisture determination techniques. Although the gravimetric method may be the most accurate method to estimate soil moisture, it can impose serious errors. In addition, it is destructive in the sense that it requires sample removal. Although this method gives the definite answer, the method is not quick or easy, but it is tedious, laborious and time consuming (Figure 5.2). In this method the water mass must be determined by drying the soil to a constant weight and the soil sample mass measured before and after drying. The water mass (or weight) is the difference between the weights of the wet and oven dry samples. The dry soil sample is the sample that has been dried to constant weight in an oven at temperatures between (100 - 110 °C), 105 °C for 24 hours is typical. Thereafter equation (5.6) was used to calculate the percentage soil moisture content in the profile at depth 0 - 100 cm.

Figure 5.2. Gravimetric method for soil moisture measurement

In this study, the soil moisture content was monitored prior to irrigation during all the seasons by using the gravimetric method (Black, 1965) from the plots of the second replication of the various treatments and the values were converted to volumetric water content by using the bulk density for each depth (Table 5.2). According to measurement of the soil water content the plots of all treatments were irrigated from the initial soil moisture content of the 0-100 cm soil layer to field capacity (FC). In this area deep percolation was assumed to be very low due to the clay soil characteristics. In contrast, groundwater table problems were also neglected because the groundwater table was at 15 m depth.

Evapotranspiration (ETc) determination by using the gravimetric method

Evapotranspiration was determined 2 - 3 days after water application from four soil samples per plot and calculated by subtracting the total soil moisture after irrigation and just before the next irrigation. The soil moisture content(θ) on a dry weight basis was calculated by using the following equation:

$$\theta = \frac{(wt\ of\ wet\ soil+can)-(wt\ of\ dry\ soil+can)}{(wt\ of\ dry\ soil+can)-\ can}*100 \tag{5.6}$$

Where:
θ = gravimetric soil moisture content (%)
Wt = weight of soil in wet or dry phases (g)

Bulk density calculation

In this study, the bulk density was calculated by using the following equations for different depths in the soil profile (secondary data), which were obtained from the linear relationship between volume weight per depth and moisture content at each depth to convert the gravimetric moisture content into volumetric content (Table 5.2) (Abdelhadi, (2006).

$$0 - 40\ cm = -0.00003*\theta^2 - 0.0019*\ \theta +1.274 \tag{5.7}$$

$$40 - 60\ cm = -0.00005*\ \theta^2 - 0.0027*\ \theta +1.386 \tag{5.8}$$

$$60 - 80\ cm = -0.0098*\ \theta +1.635 \tag{5.9}$$

$$60 - 80\ cm = -0.0145 *\ \theta +1.8517 \tag{5.10}$$

Where θ is the gravimetric soil moisture content (%).

Table 5.2. Volume weight (g/cm^3) of Gezira clay (adapted from Farbrother, 1996)

Moisture content (%)	0 – 40 (cm)	40 - 60 (cm)	60 - 80 (cm)	80 - 100 (cm)	100 - 200 (cm)
45 - 55	1.11	1.13	n.o	n.o	n.o
35 - 45	1.15	1.14	1.25	1.29	1.30
25 - 35	1.19	1.26	1.32	1.38	1.40
15 - 25	1.23	1.32	1.46	1.53	1.54
5 - 15	1.25	1.35	1.53	n.o	n.o

Note: n.o: Not observed in the field.

The evapotranspiration (ETc) during the two growing periods is shown in Table 5.3, which is strongly affected by climatic conditions. The seasonal ETc values were calculated as summation of soil moisture depletion from planting till maturity stage. It was noted that the higher air temperature and lower relative humidity resulted in high reference evapotranspiration (ETo) in the first season. Due to this the crop water needs were higher resulting as well in higher evapotranspiration (486 mm) compared to the second season. This was due to an increased amount of irrigation water (Goksoy et al., 2004). ETc variability was high in Sunflower with a low ETc of 215 mm in the second season. This also could be attributed to situation of the soil before the present season because it was fallow the in previous season and due to plot to plot variability in irrigation and crop management in the clay soil. Evapotranspiration values were more or less similar under irrigation at 10 and 15 days intervals irrespective of crop stages. The irrigation interval in the first season was every seven days, while in the second season it was a 10 days interval. It is clear that increasing soil moisture in the root zone causes a significant increase in the seasonal water consumption. These results may be attributed to the high availability of water at lower depletion of the soil moisture, which in turn increases transpiration from vegetative growth and evaporation from the soil surface.

Table 5.3. Seasonal evapotranspiration (mm) for the winter seasons 2011 and 2012

Irrigation treatment	Evapotranspiration (ETc) (mm)	
	2011	*2012*
Every week (W)	486	285
10 days after 50% flowering (F1)	412	264
15 days after 50% flowering (F2)	377	215
20 days after 50% flowering (F3)	405	208
10 days after seed filling stage (S1)	399	302
15 days after seed filling stage (S2)	398	268
20 days after seed filling stage (S3)	350	218

For crops that receive full irrigation, it is expected that ETc will be greater if the soil surface and leaves are frequently wetted. Thus the irrigation method can influence the actual evapotranspiration (ETa). The total amount of water used by the crop is proportional to the amount of water available in the soil profile. In the first year, a major part of the soil moisture was depleted from the 40 - 80 cm soil layer. The depletion from the lower layer of 80 - 100 cm ranged between 18 and 28% in pre-irrigation conditions and between 25 and 27% in post-irrigation conditions (Figure 5.3*Figure*). Among the irrigation treatments, maximum depletion of the 0 - 40 cm layer was recorded under the highest irrigation frequency (F1, S1) and weekly irrigation (W) between 33 and 36% and between 27 and 33% was recorded under prolonged irrigation intervals (Figure 5.4). Plants are thus supplied with sufficient water can transpire and meet their full ET requirements throughout their growing seasons.

Figure 5.5 depicts the actual evapotranspiration (ETa) of Sunflower obtained by the gravimetric method. The ETa was estimated on a weekly basis, which ranged from December through February. ETa was approximately 3.4 mm/day at the beginning of the season in December, gradually increased to 7.4 mm/day in the middle of the growing season in January and reached the maximum value in the middle of January. Then it reduced drastically to 2.7 mm/day at the end of the season due to leaf senescence. These results were found to be the same in both seasons. As long as sufficient water is available for irrigation and for the other constraints that limit growth, the crop water use equals the water requirements. Figure 5.6 shows the actual evapotranspiration and reference evapotranspiration, ETa values were low in the beginning of the season and increased gradually to reach maximum value in the mid season, while the ETo was high in the beginning of the season, decreased in the mid season and increased again in the late season.

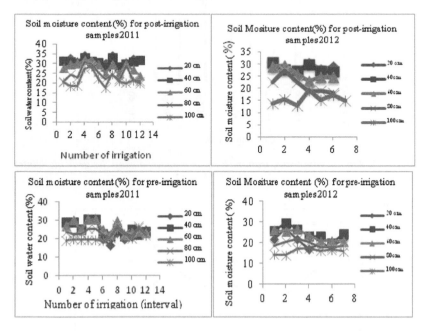

Figure 5.3. Soil moisture variability during a number of irrigations within a weekly cycle

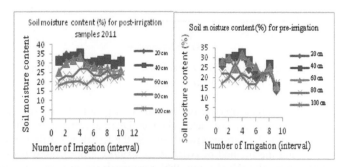

Figure 5.4. Soil moisture variability for a number of irrigations during pre-irrigation at 20 days interval

5.6 Estimation of the Winter Sunflower crop coefficient (Kc)

The crop coefficient (Kc) is a function of crop growth stages and surface soil moisture. The crop coefficient is based on experimental data, which are adjusted to reflect soil moisture changes due to irrigation or precipitation. For good irrigation planning and management purposes and for development of irrigation schedules, 10 days or monthly average crop coefficients are more effective than the Kc computed on daily basis.

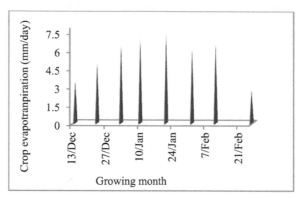

Figure 5.5. Actual evapotranspiration (ETa) in mm/day during the growing winter season 2011/2012

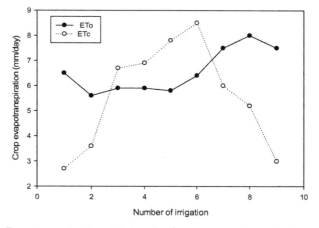

Figure 5.6. Evapotranspiration (ETc) and reference evapotranspiration (ETo) during the growing winter season 2011/2012

Soil moisture was measured before and after irrigation during the crop growth cycle, the crop height and the leaf area change. Due to consequent differences in evapotranspiration during the various growth stages, the Kc for a Sunflower varies over each period. Following the method of Allen *et al.* (1998), crop growth is divided into four development stages, the initial stage, the crop development stage, the mid season and the late season stages (Figure 5.7). The observed development stages for Winter Sunflower are given in Table 5.4. Table 5.4 shows the length of each stage obtained from the field experiments for this study accordingly to the FAO description. The total growing season for the Sunflower crop under Gezira conditions varied between 110 to 120 days.

The daily reference evapotranspiration for 2011 and 2012 was calculated with daily meteorological data by using the EToCalc software of FAO (2009). During these two years the ETo values of the 2012 season were high.

The Kc values that were adapted from this study were significant and the variation between the seasons was obvious. During the initial stage the crop coefficient varied between 0.53 and 0.59, during the development stage between 0.95 and 1.1, during the mid season stage between 1.2 and 1.3 and during the late season between 0.56 and 0.63. It was shown that the Kc varied steadily during both seasons (Table 5.5). The Kc estimation in the third season faced many problems with respect to the collection of soil moisture samples due to insufficient labourers in the research centre.

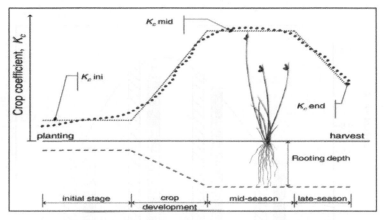

Figure 5.7. Crop growth development stages and crop coefficient
(adapted from Bos et al., 2009)

Table 5.4. The crop development stages

Crop stage	Number of days per stage (adopted)
Initial stage	22
Crop development	35
Mid season	30
Late season	25
Total	112

Table 5.5. Sunflower crop coefficient for each irrigation treatment for two growing seasons

Treatment stage	W1	F1	F2	F3	S1	S2	S3
				2011/2012			
Initial	0.53	0.53	0.50	0.50	0.59	0.55	0.53
Development	1.10	1.03	1.2	1.15	1.21	1.01	1.09
Mid	1.30	1.24	1.3	1.28	1.31	1.21	1.33
Late season	0.63	0.61	0.57	0.62	0.60	0.59	0.63
				2012/2013			
Initial	0.53	0.52	0.53	0.50	0.54	0.54	0.56
Development	1.21	1.21	1.06	0.95	1.22	1.11	1.16
Mid	1.31	1.32	1.30	1.21	1.33	1.20	1.31
Late season	0.61	0.62	0.63	0.60	0.63	0.56	0.57

Generally the crop coefficient was small in the initial stage, increased gradually with the leaf area increase during the crop development stage, reached the maximum value when crop attained maximum canopy in the mid season and decreased dramatically when the crop canopy decreased and the lower leaves became dry in the late season (Figure 5.8). On the other hand, canopy cover gradually developed in 15 - 20% between planting time (middle of July and middle of August (30 - 35) days after sowing (DAS)). In the following days in the crop development stage the canopy cover rapidly increased and reached the maximum values (80 - 85%) in the middle of September. The Kc values for different growth stages are depicted in (Figure 5.8). Generally, Kc of crop depends mainly on stage of the crop, crop type and climatic factors.

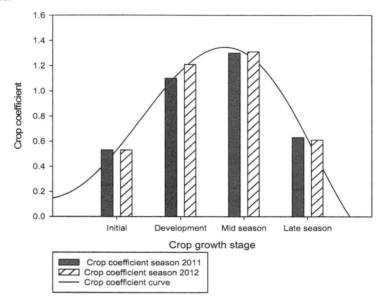

Figure 5.8. Crop coefficient (Kc) of four stages for weekly irrigation for two seasons

5.7   Concluding remarks

This Chapter demonstrates the method of crop coefficient calculation for irrigated Winter Sunflower *(Helianthus annuus L.)* crop under different irrigation regimes. The reference evapotranspiration (ETo) was calculated by using the Penman-Monteith equation and the crop coefficients for different growth stages were determined. The total growing season for the Winter Sunflower crop under Gezira conditions varied between 110 and 120 days.

The calculated ETc of Sunflower during the initial stage was low (3.4 mm/day) and increased gradually to peak values during the development and mid season stages (average of 7.4 mm/day). The maximum value of ETc was at the mid season stage. At the end of the season it decreased again to reach an average of 2.7 mm/day.

The crop coefficient for irrigated Sunflower was small at the initial stage (0.53) and increased gradually with increasing leaf area, which reached (1.2) at the development stage and reached the maximum value during the mid season (1.30) and decreased consistently to 0.63 during the maturity stage.

6 Deficit irrigation effect on Sunflower production under different environmental conditions

The major agricultural use of water is for irrigation, which is affected by the levels of water supply. Therefore, for increasing water use efficiency innovations are needed, which there are several possible approaches. In this regard, deficit irrigation is one way to maximize water productivity for higher yield per unit of water used, either during the particular stage or throughout the whole growing season. Traditional irrigation practices are differing from deficit irrigation practices. Before conducting a deficit irrigation practices, it is necessary to know crop yield responses to water stress, either during defined growth stages or throughout the whole season. This Chapter presents information on water relations and water management of Sunflower under deficit irrigation conditions and provides links to other sources of information. Moreover, this chapter reviews the effect of different irrigation water levels on yield and yield components of Sunflower under different climatic conditions.

6.1 Concept of deficit irrigation

Deficit irrigation (DI) is a water management strategy that is based on the objective of reducing irrigation water use, while improving water productivity. In many areas of the world, irrigation allocation at the farm access is less than what is required. Sometimes the cropping intensity used in the schemes is reduced as a result of the high cost of irrigation in relation to the marketing conditions of the yield of the crops. To adopt deficit irrigation needs appropriate irrigation scheduling and higher irrigation efficiency. Then the application of a water price policy would become flexible (Rodrigues and Pereira, 2009).

In arid and semi-arid regions, water scarcity is main limiting factor, where it is difficult to apply full crop water requirements for sustainable maximum growth and yield. Therefore, it is important to determine how to maintain optimum crop yields under water deficit conditions. The application of regulated water deficit strategies is one of the most promising methods to improve water productivity.

Water stress has a large impact on yield compared to the limited impact on biomass. In simple terms, crops that face water stress at critical stages may have a considerable lower harvestable yield than crops imposed to the same water stress during less critical stages. It was found that full irrigation from flowering till late flowering resulted in the highest seed yield, and also the highest oil yield and seed weight (Unger, 1983). Also the total water use increases with increased number of irrigations but yield increases were not proportional to water use, especially for adequate and full irrigation treatments.

6.2 Deficit irrigation and Sunflower yield

The yield response to water deficit of different crops is of major importance in production planning. Water deficit of crops and resulting water stress affect evapotranspiration (ETc) and crop yield. Under such conditions, water stress may develop in plants, which adversely affects crop growth and ultimately crop yield. For the evaluation of limited water supply on yield and production, consideration must be given to the effect of the limited water supply during the growing stages of each crop. The crops response to different levels of water stress needs to be known if deficit irrigation is to be successfully implemented.

Water deficit during the growing season may reduce the yield and affect yield components. In seasons when there is insufficient water for the crop demand, the optimum use of irrigation water is essential for water resources management. Optimum use implies efficient irrigation water use and proper timing of irrigation to face the critical stages of growth of the crop concerned. It is important to analyze the effect of water supply on crop yield. So if the water supply does not match the crop water requirements, the actual evapotranspiration (ETa) will be below the potential evapotranspiration (ETp). This means that water stress may develop in the plant. The effect of water stress on growth and yield depends on crop variety, magnitude and time of occurrence of the water deficit.

With respect to these aspects it is first of all of importance that Doorenbos and Kassam (1979) found that the response of yield to water supply can be quantified through a yield response factor (Ky), which relates relative yield decrease to relative ETc deficit. They have estimated the average (Ky) value for Sunflower at 0.95. It is highly important to know the (Ky) values from the literature and from those determined for particular crop species under specific sets of climatic and soil conditions.

Ghani et al. (2000) concluded that the number of filled seeds increased linearly with each increase in irrigation frequency. They found that the six irrigations gave a significantly higher number (972) of filled seeds per head than two irrigations that resulted in a minimum number (637). They summarized that the increased number of filled seeds under four and six irrigations due to the recognized role of timely and adequate availability of irrigation water. This clearly had an effect on plant growth and seed development. Beyazgül et al. (2000) found that a long period of water deficit at the sensitive growth stages causes significant reduction in seed yield. Severe water deficit during the early vegetative growth results in reduced plant height but may increase root depth. The flowering stage is the most sensitive to water deficit, which may cause considerable decrease in yield. Seed formation is the next most sensitive stage to water deficit, causing a severe reduction in both yield and oil content. In this regard, one of the most important farm management challenges is to achieve a good crop performance by providing enough water at critical growth stages.

Chimenti et al., (2002) reported that the occurrence of water stress at anthesis to physiological maturity had no significant effect on biomass but the effect on yield and harvest index was clear. Flagella et al.(2002) stated that irrigation has a highly positive effect on seed yield and that this confirms the key role of supplementary irrigation at critical growth stages, which are sensitive to water stress.

Iqbal et al (2005) stated that worldwide water stress is considered as one of the most important factors limiting crop development and yield and that several studies have been conducted to assess the effects of water stress on plant growth, yield and leaf photosynthesis.

Previous studies mentioned different effects of water deficit on Sunflower yield. In a more detailed study, Faisal et al. (2006) examined the effect of three irrigation intervals (7, 14 and 21 days) on seed and oil yield of Sunflower. They found that Sunflower was sensitive to the long irrigation intervals and that the reduction in seed and oil yield under prolonged irrigation was associated with significant reduction in the yield components. They also found that prolonged irrigation decreased the mean 1000–seeds weight. Particularly, water stress at the flowering stage reduced fertilization and seed set due to dehydration of pollen grains. More studies focused on the effect of drought on Sunflower yield. However, water stress effects on the leaf area index (LAI) showed significant difference in different days under various levels of drought on Sunflower, while daily watering produced the highest LAI at all the stages. The reduction in yield was higher in plants that received a 4 days drought cycle.

There are several publications of responses of Sunflower to different irrigation levels and water stress at different crop stages. Nazariyan *et al.* (2009) showed the effect of water stress on head formation until the end of the growing season. They found that the yield components were affected, especially the 100-seeds weight and the head diameter. They also mentioned that the drought stress severely decreased the biological yield.

Seyed (2009) examined water deficit during stem elongation and at flowering stages with two row spacing of 50 and 75 cm on growth, yield and oil yield of Sunflower. He indicated that the maximum seed yield was obtained from full irrigation (100%) and minimum seed from deficit irrigation (50%). He also found that the interaction effects of row spacing and deficit irrigation resulted in the highest seed yield and oil percentage that were obtained from full irrigation at 75 cm row spacing. He also mentioned that deficit irrigation significantly affected seed yield and oil percentage. He concluded that the highest seed yield of (2730 kg/ha) was obtained from irrigation after 50 mm evaporation as measured in a class A pan.

On the other hand, water stress and agronomical practices such as sowing date have a direct effect on Sunflower yield. Asbag *et al.* (2009) investigated five water regimes at different crop development cycles, after sowing, flower, bud initiation, complete flowering, seed filling at three different dates in Western Iran. They found that yield and yield components were positively affected by irrigation and by the earlier sowing date. They also observed that the yield and supplementary irrigation showed notable increase at the early sowing date.

Yawson *et al.* (2011) reported that Sunflower is considered to be tolerant to water deficit to some extent. In areas with low rainfall, supplementary irrigation can boost the productivity of the crop. However, it also has the ability to withstand short periods of severs soil water deficit of up to -15 atmosphere tension.

Moreover, (Abdou *et al.*, 2011) investigated three irrigation scheduling practices (according to cumulative pan evaporation (C.P.E) of 0.8, 1.0 and 1.2 and three sowing dates: the 1st and 15th of June, and the 1st of July. They indicated that irrigation scheduling and sowing date had a significant effect on seed yield and yield components. The highest seed yield was obtained from the early sowing date and irrigation scheduling at 0.8 C.P.E.

Pejić *et al.* (2013) investigated the yield response factor of Sunflower under deficit availability of water in Serbia. They found that the yield response factor (Ky) was 0.20 for the total growth period and 0.27, 0.31 and 0.48 for the vegetative, flowering and yield formation stage, respectively. They also summarized that the period from flowering to maturity was the most sensitive with respect to water deficit.

7 Field experiments and study area

7.1 Description of the study area

Gezira Research Station is located between Latitude 14° 22 to 14° 25 N and Longitude 33° 29 to 33° 30 E in Gezira State in Central Sudan. The research station is under the responsibility of the Agricultural Research Corporation, which is responsible for most of the research on agricultural problems in Sudan and has a direct collaboration with Gezira Scheme (Figure 7.1). Gezira clay plains fall between the latitudes of 13.5° and 15.5° North and the longitudes of 32.5° and 33.5° East. The soils are predominantly typical Vertisols that expand and shrink markedly with changes in moisture content and develop deep vertical drying cracks. These soils developed from deposits carried by the Blue Nile from the Ethiopian Highlands. The Gezira clay plains show a gentle slope of 0.02% from South to North. The rainy season extends from July to October with an average annual rainfall of 350 mm in the South and 150 mm in the North. The average rainfall at WadMedani is 300 mm per year. The major rainy season lies between June and October, while the minor winter season occurs from November till March, leaving April as essentially the dry season. Clay content may reach 60% or more throughout the soil profile. The content of organic carbon and nitrogen is very low (Fadl, 1971; Adam *et al.*, 1983). Some of the chemical attributes of this soil are given in (Table 7.1).

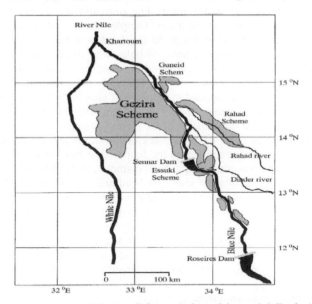

Figure 7.1. Location of Gezira Scheme (Adapted from Abdelhadi, 2000)

7.2 Climate of Gezira area

The range of climate of the central clay plains of Sudan as classified by Walsh (1988) extends from tropical continental desert in the North to tropical sub-humid in the South. Apart from the most northerly blocks of the Main Gezira, all the major large-scale developments in irrigation lie in the intervening climatic zone described as 'Tropical

Semi-dry' as cited by Farbrother (1996). Recently it was found that Gezira has three climatic zones, semi-desert in the North, dry in the centre and semi-dry in the South (Adam, 2014). The Gezira clay soil lies within the dry zone. The maximum temperature ranges from 34 °C in January to 41 - 42 °C in April and May, while the minimum temperature ranges from 14 °C in January to 25 °C in June.

7.3   Experiment One (winter season 2011, 2012 and 2013)

Treatment at different irrigation intervals

7.3.1   Experiment description and lay out

The experiments were carried out for the successive winter seasons of 2011, 2012 and 2013 in the experimental field of the Gezira Research Station Farm (GRSF) in WadMedani. The objectives were to study the effect of different irrigation intervals after the flowering and seed formation stages on Sunflower yield and yield components and to determine the water productivity under different irrigation regimes. The experiments were arranged in a randomized complete block design with three replications. Sunflower Hysun 33 (hybrid), a local adopted variety, was grown. The Crop was sown during the second week of November during the winter seasons of 2011 and 2012 and in the first week of December in the season of 2013. Three seeds were sown manually per hole on the top of the ridges, and then thinned after plant establishment (three weeks) to one plant. The row orientation was North-South to facilitate wind movement and microclimate. The plot size was 10 m in length and 5 m in width, the total area of each plot was 50 m². Each plot had 6 rows of 0.25 m top width at 80 cm apart and the space between holes was 30 cm. The distance between each plot was 2 m to avoid water movement. No runoff was allowed from the furrows by closing their ends. Weeds were controlled manually three times during the growing season. Nitrogen fertilizer was applied in the form of urea (46% N) as recommended by Agricultural Research Corporation in split doses, one with the second irrigation (with thinning) and the second at the initiation of the flowering stage.

The maturity was taken when the back of the heads was brown and the leaves were dry. Five plants were selected from the four middle rows in the plot to determine the crop parameters such as plant height, head diameter, stem thickness, yield and yield components; mainly the number of seeds per head and their weight, 100-seeds weight and seed yield. The heads were separated and dried in the open air. Plant samples were taken four times during the growing season to determine biomass and growth rate. Data were analyzed by using means of statistical analyses and means were compared with Tukey's HSD (honest significant difference) test at probability level of 5% when the effect of the treatment was significant. All the data were analyzed on the basis of the experimental design by using Statistix-9 computer software (www.statistix.com).

7.3.2   Irrigation water measurement

Water applied for each treatment was measured during the whole season by using a current meter device (Figure 7.2), which was fixed in the small canal named Abu vi (Abusitta) to measure the water flow applied to each plot. The total number of plots was 21 (Figure 7.3). The duration of water supply was calculated by determining the time (s) at the beginning of irrigation when water entered the plot and at the end of the irrigation. The difference between the two times and the velocity measured by the current meter was used to calculate the discharge (m³/s) by the following equation:

Q = V.A   (7.1)

Where:
Q   = discharge (m³/s)
V   =velocity (m/s)
A   = cross-sectional area (m)

Figure 7.2. Water measurement by using a current meter

Figure 7.3. The experimental plot shows the irrigation method

7.4   Experiment Two (summer season 2012 and 2013)

Three irrigation intervals and two intra-row plant spacing treatments

Experimental lay out description

Two experiments were conducted during the 2012 and 2013 summer seasons, Hysun 33 (Hybrid) was tested in the first experiment and Bohooth-1 (local variety) in the second experiment. In both experiments intra-row plant spacing of 30 and 40 cm assigned as S1 and S2 were applied, in addition to three irrigation intervals of 10, 15 and 20 days designated as W1, W2 and W3 respectively. Each sub-plot was 10 m in length and 5 m in width (total area 50 m²) with 6 rows. The crop was sown in the second week of July (14 July) in the first year (2012) and in the third week of July (19 July) in the second year (2013). Three seeds were sown per hole and then thinned to one plant per hole two weeks after emergence. All other agronomic practices were kept uniform for all the treatments. The amounts of water applied for each treatment were measured during the whole growing season by using a current meter, which was fixed in the inlet of Abu sitta (small canal to irrigate 2.1 ha (5 feddan)) to measure the discharge of the water entering each plot separately. The total number of plots was 18. The first irrigation was measured, but was not considered in the calculation of total seasonal water applied due to the big losses in water application because of the huge cracks, which are characteristic for the clay soil (Vertisol). Fertilizer in the form of urea at the rate of 205 kg N/ha (86 kg N/feddan) was added as recommended for Sunflower by the Agricultural Research Corporation, half with the second irrigation (after 10 days) and the second half at the flowering stage. Weeding was done three times during the crop establishment and vegetative stages. Plant samples were taken four times every 15 days, one at the initial stage and two during crop development and one at late stage of crop growth to estimate biomass. Before harvest and after the milk stage completion five plants were selected randomly from each plot to determine different crop parameters such as plant height, head diameter and stem diameter. Harvest was done at the end of October. The two inner rows were harvested to calculate the final seed yield and the above ground biomass. Table 7.1 shows the analysis of the chemical and physical properties of the soil samples taken from the experimental site, which was done in the Land and Water Research Centre Laboratory, WadMedani. The climatic data were obtained from the nearest WadMedani Meteorological Station, including: maximum and minimum temperature (^{0}C), relative humidity (%), wind speed (m/s), sunshine (hr) and rainfall (mm). Four soil samples were taken before and after each irrigation time to determine the soil moisture content for five depths at 20 cm interval, which included 20 - 40, 40 - 60, 60 - 80 and 80 - 100 cm. Table 7.2 shows the mean monthly climatic data for the study area for the first summer season. A maximum temperature of 39.5 °C was recorded in October and a minimum temperature of 18.7 °C in November. Table 7.3 gives details of amount of rainfall received during the growing season.

Table 7.1. Analysis of soil chemical and physical properties of the experimental field

Depth cm	0 - 30	30 - 60	60 - 90
Clay (%)	58	60	54
Silt (%)	25	28	31
Sand fine (%)	13	9	7
Sand coarse (%)	4	3	3
BD (g/cm^3)	1.60	1.52	1.78
FC (%)	38.2	45.9	41.9
PWP (%)	20.7	24.9	22.8
OM (%)	0.34	0.31	0.13
pH	7.9	8.1	8.0
Hydraulic conductivity (cm/hr)	2.23	2.61	0.22
CaCO$_{3\%}$	5.2	4.8	5.0
ECE (dS/m)	0.66	0.68	Na
ESP	9	15	14
CEC cmol./kgsoil	46	45	37
Av.P mg/kg soil	3.2	2.4	2.8
N%	0.050	0.045	0.040

(Source: Land and Water Research Centre Laboratory, 2012)

Table 7.2. Mean monthly climatic data of Gezira Research Station (2012)

Month	Maximum temperature (°C)	Minimum temperature (°C)	Relative humidity (%)	Sunshine (h)	Wind speed (m/s)
July	36.1	23.1	62	6.5	4.2
August	32.9	23.0	67	5.8	3.3
September	37.3	22.9	57	10.1	1.6
October	39.5	20.8	40	10.1	1.6
November	37.1	18.7	40	10.5	1.6

(Source: WadMedani Meteorological Station, 2012)

Table7.3. Daily amount of rainfall during the growing summer season 2013

Day of the month	Rainfall (mm)
15 Jul	38.0 (before planting)
18 Jul	1.7
24 Jul	2.0
2 Aug	4.8
4 Aug	2.2
6 Aug	24.3
9 Aug	23.1
11 Aug	8.0
12 Aug	19.4
15 Aug	5.6
22 Aug	26.1
23 Aug	3.7
25 Aug	4.2
26 Aug	5.0
28 Aug	0.5
30 Aug	1.5
19 Sep	6.6
Total	177

7.5   Experiment Three (winter season 2012/2013 and 2013/2014)

Three irrigation intervals and two intra-row plant spacing treatments

Experiment Three was designed as a split plot experiment, with three replications, each sub-plot was10 m of length and 5 m of width (total area 50 m^2) with 6 rows. The treatment was three irrigation intervals of 10, 15 and 20 days designated as W1, W2, W3 and intra-row plant spacing of 30 and 40 cm, specified as S1 and S2. The tested variety was Hysun 33 (Hybrid). The crop was sown in the third week of November in 2012 and in the first week of December in 2013. The total number of plots was 18. Three seeds were sown in each hole and then thinned to one plant two weeks after emergence in the first season and three weeks after emergence in the second season. All other agronomic practices were kept uniform for all the treatments. The amount of water applied for each treatment was measured during the two growing seasons by using a current meter. Weeds were controlled manually three times during the two growing seasons. Harvest was done manually during the second week of March in the first winter season 2012/2013 and during the end of March in the second season 2013/2014.

Soil samples were taken at the time of irrigation and three days after irrigation (equivalent field capacity FC) during the whole growing season to determine the soil moisture content. The samples were taken at 20 cm interval by using the gravimetric method and weighted immediately to calculate the wet weight and after drying in an oven of 105 °C for 24 hours. The dry weight was then recorded. The soil physical properties are good. The field capacity (upper limit) is 38.2 - 41.9%, the wilting point is 20.7 - 22.8% and the bulk density is 1.6 - 1.52 and 1.78 g/cm^3 for 0 - 30, 30 - 60 and 60 - 90 cm respectively. Analysis of the chemical and physical properties of the soil samples taken from the experimental site was done in the Land and Water Research Centre Laboratory (Table 7.1). According to this analysis, the pH level was normal 8%, total nitrogen 0.05%, CaCO$_3$ between 4.8 - 5.2%, but organic matter was low 0.31%. Fertilizer in the form of Urea was added during the experiment as required to maintain soil fertility. Statistical analysis was done by using the Statistix 9 software computer program. The difference of means was identified by the standard error at (P ≤ 0.05) level according to Tukey's HSD.

7.6   Crop parameters measurement

Days to 50% flowering

Days from sowing to 50% flowering were counted for each treatment to determine the effect of different treatments on it.

Days to physiological maturity

Days from sowing to physiological maturity (Figure 7.4) were counted for each plot for the different treatments. Experimental plots under full irrigation reached late to the physiological maturity stage compared to the other two irrigation treatments. That means that the duration of the flowering and seed filling stages was longer (11 days) in the full irrigation treatment. This leaded to good seed size and weight.

Figure 7.4. Start of leaves yellowing (senescence) as the indicator of physiological maturity

Crop parameters

At physiological maturity stage when the lower leaves start to dry and before heads lean, five plants were chosen randomly in each plot and tagged to measure plant height, head diameter and stem diameter, then the mean for each parameter was calculated.

Yield components

At maturity five plants were selected to determine the yield components such as number of filled seeds per head, the number of empty seeds per head and their weight, 100-seeds weight (g) and total seed yield (kg/ha). Filled seeds per head were sorted out and their numbers to total seed were computed. Also 100-seeds weight was randomly taken from each harvested yield from each plot and the mean of 100-seeds weight was determined.

8 Results and discussion of the field experiments

The increasing demand for food and other agricultural products makes the conservation of water resources increasingly imperative. This is more so in arid and semi-arid regions, which have a shortage in water supply during the peak of the growing season, due to irrigation mismanagement. The agricultural sector in Sudan has a low output. However, the low output of the sector has been attributed to inadequate inputs, untimely field operations, inefficient crop production and irrigation technologies. In this Chapter the findings of the three experiments, conducted under different irrigation management to test deficit irrigation strategies under Gezira climatic conditions, will be presented.

8.1 Different irrigation intervals (experiment One)

Effect of weather on crop

High air temperatures recorded during the seed development and maturity stages had a significant effect. A reduction in seed yield under the 20 days irrigation interval was variable due to confounding environmental factors such as water stress. Effects of air temperature during the crop development stage were not observed but the higher temperature during the flowering and seed filling stages was recorded and may have accelerated the physiological maturity (Figure 8.1) (Canavar *et al.*, 2010).

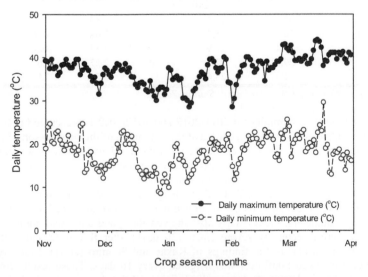

Figure 8.1. Daily maximum and minimum temperature (°C) for two winter seasons (2011 and 2012)

Table 8.1 displays the weather data for two growing seasons. The average of air temperature over the two seasons, February was the hottest (37.0 °C) followed by November (36.6 °C), and December (34.6 °C) respectively. The warmest January was that of 2012/2013 and the coolest of 2011/2012. The relative humidity generally decreased from November throughout to February. While ETo was higher in January and February in 2012 season (6.1 and 7.6 mm/day) than in 2013 season (5.8 and 6.8 mm/day). Higher ETo was recorded during the physiological maturity periods for the three seasons. This may affect the seed development and accelerates the physiological maturity period. Soil water depletion increases when ETo increases, thus crops are sensitive to days of low or high ETo.

Table 8.1. Mean monthly weather data of Gezira Research Station for the three growing winter periods (2011, 2012 and 2013)

Month	Maximum temperature (°C)	Minimum temperature (°C)	Relative humidity (%)	Wind speed (m/s)	Sunshine (hour)	ETo (mm)
2011/2012						
November	36.1	15.5	35	1.9	11.2	6.0
December	35.3	15.2	31	1.9	10.9	5.6
January	33.8	15.3	31	2.3	10.5	6.1
February	38.4	19.1	28	2.5	10.3	7.6
March	38.9	19.8	24	2.5	9.9	8.0
2012/2013						
November	37.2	18.6	40	1.7	10.3	5.6
December	34.8	15.9	31	2.8	10.7	5.4
January	34.8	17.3	33	2.8	10.2	5.8
February	37.0	20.1	27	2.3	10.2	6.8
March	40.5	19.3	23	1.8	10.4	7.2
2013/2014						
December	33.6	16.5	32	1.89	10.7	5.8
January	34.2	15.5	35	2.68	10.7	5.8
February	35.4	20.9	34	2.32	10.5	6.4
March	39.5	22.3	30	2.27	9.6	7.2
April	41.3	25.6	31	3.52	7.5	8.1

(Source: WadMedani Meteorological Station, 2014)

Daily relative humidity for 2011/2012 and 2012/2013 is presented in Figure 8.2. The high relative humidity was recorded at the beginning of the both seasons, but it was higher in the second season (2012). On the other hand, relative humidity was low at the end of both seasons (March). There was no environmental impact was observed due to high relative humidity during the two growing seasons.

Table 8.2 shows the effect of the irrigation intervals on crop yield and its components. Statistical analysis demonstrated that the water stress had a negative effect on all crop parameters. Field results also showed that yield and yield components for Sunflower were affected by severe stress (20 days) after flowering and seed filling stages. This is similar to the findings of Kakar and Soomro (2001). Moreover, seed yield increased when Sunflower was irrigated every 10 days. These results emphasize the importance of adequate water supply during seed development to obtain high seed yield. As proved by this field research full and limited irrigation applied during flowering and seed filling stages significantly increased seed yield. This is similar to the findings by Erdem *et al.* (2001). They indicated that full irrigation during flowering and moderate stress during seed filling stages significantly increased seed yield.

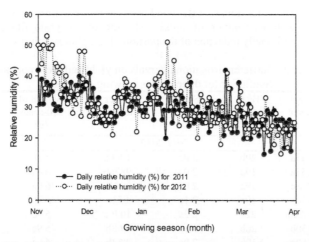

Figure 8.2. Daily relative humidity (%) for two winter seasons (2011 and 2012)

In this study there were significant differences between the irrigation treatments, the highest number was recorded at weekly irrigation and the lowest at 20 days interval after the flowering stage (F3). The highest seed yields obtained from weekly irrigation were 3130, 3140 and 3100 kg/ha for the first, second and third season respectively (Figure 8.3). On the other hand, the lowest yields that were obtained under irrigation at 20 days interval after the flowering stage (F3) were 2080, 2130 and 2260 kg/ha in the first, second and third season respectively. Increased irrigation frequency increased the number of filled seeds and thus seed yield. Karam *et al.* (2007) reported that the increase in the irrigation interval reduced seed yield, plant height, head diameter; oil seed content and increased the number of un-filled seeds. The higher seed yield resulted principally from an increase in the number of harvestable seeds. However, they did not find significant increases in seed yield by increasing the irrigation period to physiological maturity.

There is a direct relationship between number of irrigations and seed yield. Yagoub *et al.* (2010) found that irrigation 7 days increased Sunflower seed yield by 43% more than 21 days irrigation interval. The highest seed yield was obtained from increased number of irrigations due to the increase in available soil moisture in the root zone as reported by Rawson and Turner (1982) and Kakar and Soomro (2001). Sunflower variety Hysun 33 gave the highest seed yield after receiving irrigation at the 7 days interval, starting 40 days after sowing (Stone *et al.*, 1996). They found that seed yield increased when Sunflower received full irrigation after the flowering stage. Severe stress during flowering and seed filling resulted in more empty seeds (Human *et al.*, 1990). On the other hand Asbagh *et al.* (2009) showed that frequent irrigation resulted in the highest seed number per head. There is a curvilinear relationship between seed number per head and plant growth rate in Sunflower (Vega *et al.*, 2001).

Under water stress conditions, plants do not absorb enough water and hence, the seeds are more or less unfilled. Results showed that when water stress occurred, particularly during flowering, the head diameter decreased and as a result the number of filled grains decreased. However, water stress reduced the seed number by reducing the leaf area. This may have attributed to the negative impact on the transport of photosynthesic materials. Roshdi *et al.* (2006) concluded that water stress during the flowering stage has a significant effect on yield and yield components of Sunflower.

Reducing seed yield under limited water conditions could be contributing to the effect of water deficit with acceleration of aging, reduction of the grain filling stage and induced stomata, and finally reduction of the portion of net photosynthesis.

Table 8.2. Effect of different irrigation intervals on yield and yield components of Sunflower for three winter growing seasons

Irrigation treatment	Plant height (cm)	Head diameter (cm)	Stem diameter (cm)	Number of filled seeds	Weight of 100-seeds (g)	Yield (kg/ha)
			2011/2012			
W	153a	20b	2.0a	1060a	5.8b	3130a
F1	153a	19c	2.0a	987ab	6.4a	2670ab
F2	148c	19c	2.0a	968ab	6.0a	2410b
F3	150b	19c	2.1a	851b	5.7b	2080c
S1	152ab	22a	2.2a	1070a	5.5c	2800ab
S2	150b	20b	2.1a	868b	5.9a	2350b
S3	150b	19b	2.0a	883b	6.0a	2190c
Mean	151	20	2.1	956	5.9	2400
SE±	2.40	0.82	0.11	102	0.38	223
CV%	2.74	2.27	9.06	18.8	11.2	16.1
			2012/2013			
W	161a	19a	1.4a	1030a	4.7a	3140a
F1	160a	17ab	1.4a	1020a	4.8a	2880ab
F2	160a	16ab	1.4a	905b	4.5ab	2510b
F3	160a	15b	1.4a	862c	4.3c	2130bc
S1	161a	17ab	1.5a	955ab	4.7a	2840ab
S2	158a	15b	1.3b	920ab	4.6b	2350c
S3	159a	15b	1.2b	880c	4.6b	2270bc
Mean	160	16	1.4	938	4.6	2590
SE±	1.43	1.23	0.09	55.2	0.18	202
CV%	1.54	13.1	12.6	10.1	6.91	10.1
			2013/2014			
W	155a	23a	1.2a	1340a	5.2a	3100a
F1	153a	21a	1.1a	1270ab	5.2a	3070a
F2	150a	18ab	1.1a	965bc	5.3a	2670ab
F3	147a	18	1.0a	908c	5.0ab	2260b
S1	154a	21a	1.2a	1060abc	4.8b	2970a
S2	155a	17b	1.0a	900bc	4.9ab	2490ab
S3	149a	18ab	1.1a	915bc	5.0a	2400ab
Mean	152	19	1.1	1052	5.1	2710
SE±	6.1	10.7	10.0	11.2	11.8	129
CV%	6.94	1.20	0.06	67.0	0.34	8.1

Results of this research showed that with increasing irrigation intervals and applying water stress, the head diameter and seed yield decreased. This confirms the findings by Roshdi *et al.,* (2006) and Ali and Talukder (2008). Largest head diameter was recorded under weekly irrigation (23 cm) and 10 days (21 cm) after the flowering and seed formation stages respectively in the third season. There was a linear relationship between head diameter and irrigation level (Taha *et al.,* 2001).

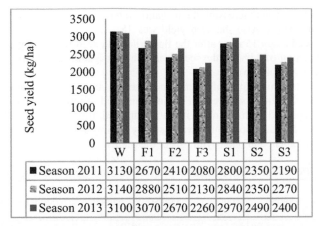

	W	F1	F2	F3	S1	S2	S3
■ Season 2011	3130	2670	2410	2080	2800	2350	2190
▨ Season 2012	3140	2880	2510	2130	2840	2350	2270
■ Season 2013	3100	3070	2670	2260	2970	2490	2400

Figure 8.3. Total seed yield (kg/ha) obtained from different irrigation treatments for the three winter seasons 2011, 2012 and 2013

Water stress during the flowering stage dramatically decreased seed yield more than water stress during seed filling stage (F3 and S3 respectively). Usually, the highest seed yield resulted from irrigation during flowering or late flowering (Unger, 1983). On the other hand, Mohammed *et al.* (2013) indicated that water stress during the flowering and seed filling stages causes considerable reduction in seed yield of Sunflower.

Limited water supply has a significant effect on crop parameters as reported by Razi and Assad (1999). Moreover, deficit irrigation at the early seed formation stage was found to slightly increase seed yield in S2 for the three seasons. The results showed that water deficit at the flowering stage needs to be avoided, while at seed formation it is somewhat acceptable as concluded by Karam *et al.*, (2007) and Mobasser and Tavassoli (2013). They reported that water stress at different crop development stages can affect final seed yield and oil content. Talha and Osman (1975) found that soil water stress during slow elongation, rapid elongation, flowering and ripening stages significantly decreased the 100-seeds weight and oil content. Because during these stages the most important yield components (seed number, and seed weight) are formed. The increase in yield was mainly due to the increase in number of seeds per head and not to the seed weight. Sunflower showed slightly less performance under irrigation at 15 days interval compared to the full irrigation treatments. However, according to previous studies the initial growth stage is the most important for adequate water supply than later irrigations. However, irrigation deficit at late growth stages of Sunflower has less effect on yield than at early growth stages (Mehmet, 2011). It is noteworthy that in Gezira Scheme farmers use irrigation scheduling for other crops to irrigate Sunflower.

8.1.1 Seed yield (kg/ha)

Seed yield was determined under different irrigation intervals to investigate the effect of different water application during the various growth stages on final seed yield. Table 8.3 demonstrates the results of seed yield and reduction percentages due to different irrigation intervals from three growing seasons. The results revealed that the highest seed yield was obtained from irrigation every week and the lowest seed yield under the 20 days interval during the flowering and seed filling stages and the other treatments varied between these two. The results indicate that reduction in seed yield was not significant 5 - 15% when increasing the irrigation interval to 10 days after the flowering and seed filling stages. Also the seed yield was 18 - 25% lower when the Sunflower crop was irrigated every 15 days after the flowering and seed filling stages compared to the higher reduction of 34% on the seed yield when irrigation at 20 days interval after the flowering stage was applied.

The results show that Sunflower is more sensitive to soil water stress during the elongation and flowering stages than during the seed filling stage. Seed yield was significantly improved with optimum irrigation applied after the start of the flowering and seed filling stages. Also enough water supply during the vegetative stage caused a good leaf development and photosynthesis of the plants.

Table 8.3. Effect of different irrigation intervals on seed yield and relative reduction

Treatment	Yield (kg/ha)	Yield reduction %	Yield (kg/ha)	Yield reduction %	Yield (kg/ha)	Yield reduction %
		2011		2012		2013
W	3130	0	3140	0	3100	0
F1	2670	15	2880	8	3070	5
F2	2410	23	2510	20	2670	18
F3	2080	34	2130	32	2260	31
S1	2800	10	2840	9	2970	9
S2	2350	25	2350	25	2490	24
S3	2190	30	2270	28	2400	26

Deficit irrigation that the irrigation application below the full ET requirement. Deficit irrigation practices can be an alternate option for improving irrigation schedules and thereby increase crop production under restricted water resources in irrigated agriculture. However, deficit irrigation at the late seed formation stage slightly increased seed yield in (S2) in comparison with the early flowering stage (Göksoy et al., 2004). Seed yield and number of filled seeds were found to vary significantly. In addition, irrigation deficiency can adversely affect the activities of reproductive organs such as grains and heads because of the high sensitivity of Sunflower to water stress during flowering and pollination stages.

Water stress during the flowering and seed formation stages may affect seed yield of Sunflower due to its effect on reproductive organs and the increase in the number of empty seeds. Previous studies reported the reduction of seed yield due to water stress (Igbal et al., 2005; Hamid and Abolfazl, 2013) Sunflower has the ability to extract water from deeper soil layers when the crop is stressed early at the vegetative stage (D'Andria et al., 1995). Accordingly, the results revealed that the water stress during flowering and seed filling stages resulted in a considerable decrease in seed yield by 30 - 34% and by 26 - 31% in the first and second season respectively. Exposure of plants to water deficit at sensitive stages led to noticeable decreases in seed yield.

8.1.2 Oil content (%)

The oil percentage is an important evaluation parameter of Sunflower quality, which may be affected by deficit irrigation. Sunflower seeds contain good quality oil (37 - 42%) as well as high amounts of protein (14.8%). Previous studies reported that the percentage of oil content of Sunflower slightly decreased when the crop was exposed to water stress at the flowering stage (Ali, 2013; Hamid, 2013). There were no significant differences observed in oil content among different irrigation treatments (Table 8.4).

The highest oil percentage (42%) was recorded with full irrigation during the whole growing season, and the lowest percentage of 37% when plants were subjected to water stress at the flowering stage, while water stress after the seed filling stage had no significant effect. Because stressing on reproductive processes, such as flowering and pollination, causes a reduction in grain yield and accordingly reduced oil yield. Oil content increased with increasing the amount of irrigation (Bashir and Mohamed, 2014). Results from two seasons clearly showed that there were no significant differences in protein content among all treatments. The results also showed that, water stress significantly (P ≤ 0.05) decreased seed yield, yield components and seed oil content. Protein content in all irrigation treatments did not show any significant difference.

Table 8.4. Oil and protein content (%) as affected by irrigation intervals for two winter seasons (2011 and 2012)

Irrigation treatment	Oil content (%)	Protein content (%)	Oil content (%)	Protein content (%)
	2011		2012	
Weekly irrigation (W)	42	14.8	41	13.9
10 days after 50% flowering (F1)	42	13.9	42	13.8
15 days after 50% flowering (F2)	41	13.8	39	14.0
20 days after 50% flowering (F3)	37	14.1	38	13.7
10 days after seed filling stage (S1)	40	14.6	41	14.4
15 days after seed filling stage (S2)	39	14.7	38	13.8
20 days after seed filling stage (S3)	38	14.2	38	14.4

8.1.3 Application of irrigation water

In Gezira Scheme the farmers irrigate their crops depending on their observations and availability of water in their canals. This caused water shortage in some areas during the growing season, even at the critical stages of most grown crops. In general the farmers tend to over-irrigate their crops, which reduces water productivity and net income. In this study furrow irrigation and famer's traditional method was used to irrigate the experimental field plots. The total number of irrigations for W, F1, F2, F3, S1, S2, S3 were 12, 11, 11, 10, 11, 11, 10 and 10, 9, 9, 9, 10, 9, 9 and 10, 8, 7,8, 8,7, 7 in the first, the second and in the third seasons respectively. Despite that the canals of Gezira Scheme were designed based on a figure of 952 m^3/ha/irrigation (400 m^3/fed/irrigation) (Abdelhadi and Elhadi, 2008) water during different irrigation intervals has exceeded this amount. First irrigation for weekly irrigation was 868 and 965 m^3/ha/irrigation in the first season and second season respectively. The average irrigation rate for W, F1, F2, F3, S1, S2, S3 was 995, 1020, 1040, 1040, 965, 987, 1030 m^3/ha/irrigation in the first season and the corresponding values were 905, 893, 822, 926, 874, 844, 838 m^3/ha/irrigation and 975, 1100, 1070, 1130, 1080, 1070 and 1090 m^3/ha/irrigation in the second and third (Table 8.5). Figure 8.4 illustrates the comparison between the seasonal water application for the three growing seasons; higher amount of water applied was recorded in the first season compared with the other two growing seasons.

In clay soils, intermittent drying may lead to shrinkage and cracking of the soil, thereby allowing to increase soil moisture loss, increase in water requirement and decrease in water productivity (high amount of water applied with low yield under irrigation at 20 days interval). Table 8.5 shows the total water applied during the three growing seasons, which illustrates that the large amount of irrigation water was in the first season due to high ETo.

Table 8.5. Seasonal water applied (m³/ha) for the three winter growing seasons for each irrigation treatment

Irrigation Treatment	2011/2012			2012/2013			2013/2014		
	Applied water (m³/ha)	Number of irrigations	Average per irrigation (m³/ha)	Applied water (m³/ha)	Number of irrigations	Average per irrigation (m³/ha)	Applied water (m³/ha)	Number of irrigations	Average per irrigation (m³/ha)
W	12000	12	995	9350	10	935	9750	10	975
F1	11200	11	1020	9400	9	1040	8810	8	1100
F2	10400	10	1040	8600	8	1080	7500	7	1070
F3	10400	10	1040	8240	8	1030	9030	8	1130
S1	10600	11	965	9480	9	1050	8660	8	1080
S2	10900	11	987	8540	8	1060	7510	7	1070
S3	10300	10	1030	8470	8	1060	7600	7	1090

The differences between the three seasons as shown in Table 8.5 could be attributed to the fact that the field location may influence the amount of water applied and to the proper land preparation. Many studies reported that the field location was influenced by the amount of water applied from head to tail of the canal to the field in Gezira Scheme (Omer *et al.*, 1991).

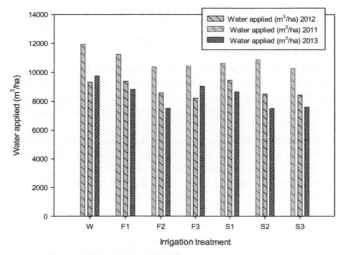

Figure 8.4. Seasonal water application (m³/ha) for the three winter growing seasons (2011, 2012 and 2013) for each irrigation treatment

8.2   Concluding remarks on experiment One

The study demonstrated the influence of different water supply levels and deficit irrigation as having the main effect on seed yield and other crop parameters and eventually on oil seed content. The following conclusions were obtained:

The irrigation interval of seven days (W) gave the highest seed yield (3130, 3140 and 3100 kg/ha) and the 20 days interval after the flowering stage (F3) gave the lowest seed yield (2080, 2130 and 2260 kg/ha) obtained from the first, second and the third seasons respectively.

Results revealed that prolonged irrigation intervals after the flowering stage had more yield reduction impacts (34%) compared with a reduction of (28%) as a result of water stress after the seed filling stage. These results suggest that a mild water deficit during the seed filling stage had little effect on seed yield.

The irrigation interval of 15 days had an insignificant impact on Sunflower seed yield and percentage of oil content. Thus, when there is shortage in water supply alternative irrigation scheduling at 15 days interval is recommended.

The irrigation treatment at 10 days interval after the flowering and seed filling stages gave the highest seed yield in all three seasons, but in the third season the seed yield was higher about 3070 and 2970 kg/ha respectively.

The high oil content (42%) was recorded under weekly irrigation and the lowest percentage was (37%) under the irrigation interval of 20 days after flowering (F3). The protein content recorded an average of 14% in all irrigation treatments. The impact of deficit irrigation on oil and protein content was insignificant.

8.3   Experiment Two (summer seasons 2012 and 2013)

Effect of intra-row plant spacing and irrigation interval on Sunflower

In the seasons where there is insufficient water for the crop demand, the optimum use of irrigation water is essential for water resources management. Optimum use implies efficient irrigation water use and proper timing of irrigation so as to conform to critical stages of growth of the crop concerned. It is therefore important to analyze the effect of water supply on crop yield. So if the water supply does not match the crop water requirements (CWR) the actual evapotranspiration (ETa) will be below the potential evapotranspiration (ETp), which means that the plant will be under water stress. The effect of water stress on growth and yield depends on the crop variety, as well as on magnitude and time of occurrence of water deficit. Hence, in order to optimize the crop water requirement, critical crop growth stages and irrigation schedules for maximizing production is highly desirable.

These experiments were conducted under rainy season conditions for two seasons 2012 and 2013. The effect of rainfall was significant on the crop development cycle because most of the heavy rainfall was occurring before the planting date and/or during the flowering stage. Figure 8.5 illustrates that the total amount of rainfall during the growing season was 123 mm. Amounts of rainfall of 17.4 and 16.2 mm were received 7 and 10 days after the planting date respectively. Also an amount of rainfall of 37.6 mm was received close to the planting date (before one week). These amounts of rainfall interrupted the days of irrigation and some of the agronomical practices such as weeding and thinning, but they were good for crop establishment and development.

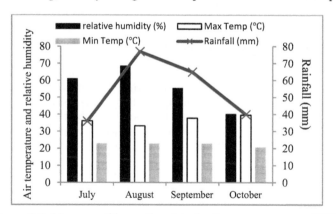

Figure 8.5. Mean monthly weather data for the summer season (2012)

During the second season the rainfall was recorded for the whole growing season. Figure 8.6 shows that the accumulated rainfall was 177 mm, but 39.7 mm were recorded in the first week of July before the planting date. The total amount of rainfall recorded during the specific days of the critical stage was 137 mm. The lower amount of rainfall (3.7 mm) was received at the end of July and a higher amount was received in August in both seasons and was distributed well, but in the second season the rainfall was higher (128 mm) and the rest (6.6 mm) was received at the end of September. Supplemental irrigation was applied to all the treatments due to insufficient rainfall to avoid water stress during the crop growth stages in order to achieve stable yields and improve water productivity.

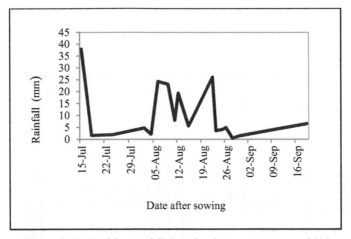

Figure 8.6. Monthly rainfall data for the summer season 2013

The effect of intra-row plant spacing on each irrigation interval was significantly different at the 5% level of probability (P ≤ 0. 05). Tables 8.6 and 8.7 show the yield and yield components for the two summer seasons 2012 and 2013. Yield reduction was remarkable in the 2013 season due to the infestation by Pashnoda insects attacking some plots during the flowering and seed filling stages and we covered the plants to avoid losses in yield (Figure 8.7). Hence, big damage was obvious in the outer ridges and erroneous values in some plots were not taken into account in the statistical analysis. Bohooth-1 (local variety) was more sensitive to the insect attacks compared to Hysun 33 (Hybrid). Compared to W1 the reduction in seed yield was 28% and 41% in W2 and W3 respectively. In fact, severe water stress W3 occurred after the flowering stage and drastically reduced seed yield. The most sensitive stages for water stress in Sunflower were before and after flowering, which are found in this study.

There were significant differences among the irrigation treatments in both seasons. Significantly increased seed yield, associated with an increase of the number of filled seeds per head. Likewise, increasing intra-row plant spacing had an incremented effect on plant height. It slightly affected stem diameter and head diameter. The plant height was reduced by 3% and 13% under irrigation at 15 and 20 days intervals compared to irrigation at 10 days interval. While in the first season irrigation at 20 days interval decreased the number of seeds per head by 20% and 21% for Hysun33 and Bohooth-1 respectively. In the second season the reduction was a largely significant difference in number of seeds per head. The reduction in Hysun 33 was higher by 18% and 22% under irrigation at 15 and 20 days intervals compared to the 10 days interval. While in Bohooth-1 it was 13% and 17% in the corresponding treatments (Tables 8.6 and 8.7). Deficit irrigation affected all aspects of crop growth and yield attributes adversely, although there were no statistically significant differences among the stressed treatments. However, the effect of intra-row plant spacing was significantly different in the average number of seeds per head over all the three irrigation treatments. The results indicate that the head diameter increased under lower densities, which was attributed to the increased seed head. This finally resulted in increase in yield as was also concluded by Ardakani (2006). Moreover, increasing water application resulted in increased head diameter as was also reported by Hajhassani-Asl (2009).

Table 8.6. Interaction effects of irrigation intervals, intra-row plant spacing on yield and yield components of Sunflower for two varieties (summer season 2012)

Irrigation /plant spacing treatment	Plant height (cm)	Head diameter (cm)	Stem diameter (cm)	Number of filled seeds	Weight of filled seeds (g)	Total yield (kg/ha)
Hysun 33						
W1S1	173a	27a	2.2a	1730a	177a	3180a
W1S2	172a	29a	2.2a	1650ab	178a	3240a
W2S1	170a	28a	2.3a	1630ab	170a	2700ab
W2S2	167ab	27a	2.3a	1560b	143b	2330b
W3S1	158b	26a	2.2a	1410c	140b	1840c
W3S2	151b	25a	2.2a	1380c	145b	1820c
SE±	2.20	0.65	0.05	28.6	3.68	40.5
CV%	2.30	4.22	3.80	3.17	13.3	2.77
Bohooth-1						
W1S1	172a	29a	2.0a	1670a	112ab	3200a
W1S2	170b	30 a	1.9a	1610a	114ab	3180a
W2S1	170b	29 a	2.1a	1450ab	126a	2630ab
W2S2	167c	29 a	2.0a	1450ab	101b	2240b
W3S1	167c	27a	2.2a	1250b	115ab	1800c
W3S2	166c	29.0a	2.1a	1310b	110ab	1800c
SE±	2.36	0.95	0.102	30.9	6.55	57.7
CV%	2.43	5.69	8.63	3.68	9.90	4.82

Means followed by different letters are significantly different at (P ≤ 0.05) level according to Tukey's HSD

Table 8.7. Interaction effects of irrigation intervals, intra- row plant spacing on yield and yield components of Sunflower for two varieties (summer season 2013)

Irrigation / plant spacing treatment	Plant height (cm)	Head diameter (cm)	Stem diameter (cm)	Number of filled seeds	Weight of filled seeds (g)	Weight of 100-seeds (g)	Total yield (kg/ha)
Hysun 33							
W1S1	180a	21a	1.6a	871a	59.0ab	14.0a	3100a
W1S2	173b	23a	1.5a	833a	64.7 a	13.7ab	2920a
W2S1	172a	21a	1.5a	791ab	56.0 b	14.0a	2500b
W2S2	171b	22a	1.5a	794ab	60.0ab	13.7ab	2200b
W3S1	168c	20 a	1.4a	596c	48.0c	13.3b	1780c
W3S2	165c	20a	1.4a	679b	48.0c	13.7ab	1870c
SE±	0.97	0.27	0.06	37.6	3.70	0.23	118
CV%	0.99	2.23	7.45	8.56	11.4	2.98	8.19
Bohooth-1							
W1S1	158b	18a	1.3a	755ab	39c	14a	3080a
W1S2	160a	18a	1.3a	807a	42ab	13a	2940a
W2S1	161a	17a	1.2a	753ab	43a	13a	2140b
W2S2	162a	17a	1.3a	700b	41b	13a	2230b
W3S1	154c	16a	1.3a	669c	41b	14a	1900c
W3S2	153c	17a	1.3a	701b	43a	13a	1900c
SE±	1.77	0.45	0.09	45.4	1.56	0.30	73.3
CV%	1.94	4.51	11.57	10.8	6.50	3.94	5.37

Means followed by different letters are significantly different at ($P \leq 0.05$) level according to Tukey's HSD

Figure 8.7. Damage caused by Pashnoda insect during summer 2013

The results showed that there were slightly significant effects of three irrigation levels regarding the crop parameters weight of 100-seeds, head and stem diameters, while significant differences were observed in the seed yield due to insufficient amount of rainfall received during the critical crop stages. Water deficit during seed development reduces the duration of seed filling, and thus, the final seed weight.

The differential response of Sunflower to deficit irrigation levels was already observed by El Naim and Ahmed (2010). In general, the differences in seed yield and other parameters among the different irrigation treatments were small, but there was a sharp contrast between irrigated and stressed treatments. The results show that there were no significant differences between the two varieties in obtained seed yield in both seasons. This result confirmed the findings of Amir and Khalifa (1991), which they found that the differences in seed yield between open pollination cultivars and hybrids under irrigation were not significant. Time occurrence can cause canopy senescence with subsequent reduction in seed yield. There is genetic variability in the response of sunflower genotypes to water deficits. Long-season genotypes have greater canopy cover and produce more biomass under drought conditions, because of their ability to extract more water from the subsoil.

Sunflower suffered severe water stress during the reproductive stages, which dramatically lowered seeds number and seed weight (Table 8.6). Decreasing seed weight under water stress is because of reduction of the grain filling stage and premature aging. In fact, Hysun 33 cultivar had a high potential in production of seed yield under optimum conditions. Adequate soil moisture during reproductive processes, such as flowering and pollination, causes the highest seed yield. Severe water stress that occurred after the crop reached the end of heading and beginning of flowering had induced a fast senescence when compared with the other treatments. Consequently a rapid decline in canopy cover was observed. This study confirms that in the cultivation of Sunflower prolonged irrigation intervals should be avoided.

Stressing the crop up to 20 days irrigation intervals reduced seed yield by 40 and 44% for Hysun 33 and between 38 and 44% for Bohooth-1 in the first and second season respectively. While an increased irrigation interval to 15 days did not result in a significant reduction in seed yield (16 - 31%). However, deficit irrigation or mild water stress at early seed formation was found to increase slightly seed yield in the W2 treatment. The results revealed that the differences between the 10 and 15 days irrigation interval in seed yield were not significantly different. Thus, shifting the irrigation interval to 15 days when there is a shortage in water supply during the growing season is recommended (Table 8.8).

In general the water applied to the crops was the same or seemed congruent in both seasons (Table 8.9). It can be noted that the general trend of the amount of applied water over the whole growing season was higher under irrigation at 20 days interval and lower under irrigation at 10 days interval due to the influence of the cracks. The differences in water amount were mainly due to the watering method (traditional farmer's method). As a direct result of bad land preparation (ploughing and levelling), the experimental fields had lost their design slope, with the yardstick (observation method) method used for crop irrigation. However, the lower parts of the field received more water than the higher parts to be satisfied at one time. The average amount of water applied in W2 and W3 was increased by respectively 6% and 11% compared to W1. However, the contribution of rain in the water supply was 18% in W1, 24% in W2 and 24% in W3. Only a part of this rain could satisfy the crop water demand at the initial stage (crop establishment). Table 8.9 showed that most of the water used was from supplementary irrigation, because the seasonal rainfall was 137 mm. The relationship between water applied and seed yield has been plotted to evaluate the effects of different water supply levels on yield loss. Figure 8.8 shows the linear relationship between water application and seed yield in both seasons. The good correlation ($r^2 = 0.82$) between water applied and seed yield indicates that increase in seed yield was associated with increase in water applied.

Table 8.8. Seed yield and relative reduction for two Sunflower hybrids during the two summer seasons

Irrigation/plant spacing treatment	Total seed yield (kg/ha)	Yield reduction (%)	Total seed yield (kg/ha)	Yield reduction (%)
Hysun 33	2012		2013	
W1S1	3180	2	3100	0
W1S2	3240	0	2920	6
W2S1	2700	16	2500	15
W2S2	2330	28	2200	29
W3S1	1840	43	1780	43
W3S2	1820	44	1870	40
Bohooth -1				
W1S1	3200	0	3080	0
W1S2	3180	1	2940	5
W2S1	2630	18	2140	31
W2S2	2240	30	2230	28
W3S1	1800	44	1900	38
W3S2	1800	44	1900	38

Table 8.9. Total water applied (m^3/ha) for Hysun 33, for the two summer seasons 2012 and 2013

Irrigation /plant spacing treatment	Total water applied+ rainfall (m^3/ha)	Number of irrigation	Average of each irrigation (m^3/ha)	Total water applied+ rainfall (m^3/ha)	Number of irrigation	Average of each irrigation (m^3/ha)
	2012			2013		
W1S1	7730	7	1104	7820	7	1117
W1S2	7740	7	1106	7790	7	1112
W2S1	6640	6	1107	5660	5	1130
W2S2	6620	6	1103	5700	5	1140
W3S1	5660	5	1132	5800	4	1450
W3S2	5610	5	1122	5830	4	1460

Table 8.10 shows the total amount of water applied and the average per irrigation during the two growing seasons. The reasons for the higher amount of water in treatment W3 compared to the treatments W1 and W2 are the higher evaporation from the soil, the wider intra-row plant spacing and the cracks in the soil under the long irrigation interval, while the evaporation from the narrow space is less. However, W1 with both intra-row plant spacing treatments received 7 irrigations, W2 (S1and S2) and W3 (S1and S2) treatments received 6 and 5 irrigations respectively. Figure 8.9 shows the results of the seasonal water applied between the three irrigation treatments, which indicated that the W1 and W3 received more and less irrigation water respectively. However, there were no significant differences among the intra-row plant spacing in seasonal water applied. Water applied increased by the increased number of irrigation, but was not affected by increases in intra-row plant spacing. These results confirmed the findings of Lamm *et al.* (2011).

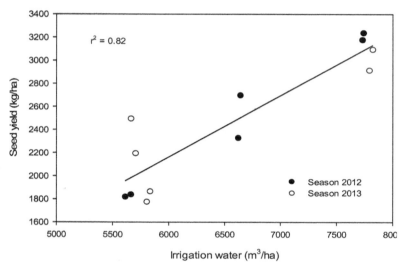

Figure 8.8. Trend of water applied versus seed yield (summer 2012/2013 season)

Table 8.10. Total water applied (m³/ha) for Bohooth-1 (local variety), for the two summer seasons 2012 and 2013

Irrigation /plant spacing treatment	Total water applied + rainfall (m³/ha)	Number of irrigations	Average of each irrigation treatment (m³/ha)	Total water applied + rainfall (m³/ha)	Number of irrigations	Average of each irrigation (m³/ha)
		2012			2013	
W1S1	7750	7	1110	7440	7	1060
W1S2	7710	7	1100	7550	7	1080
W2S1	6670	6	1110	5760	5	1150
W2S2	6750	6	1130	5780	5	1160
W3S1	5610	5	1120	5800	4	1450
W3S2	5540	5	1110	5780	4	1450

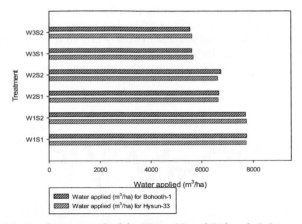

Figure 8.9. Total water applied for Hysun33 and Bohooth-1 (summer 2012)

8.4   Experiment Three (winter seasons 2012/2013 and 2013/2014)

Three irrigation intervals with two intra-row plant spacing treatments

The agronomical practices involved in Sunflower husbandry (genotype, row spacing, irrigation, fertilization) affect yield by developing crop growth and biomass. Among these agronomical practices, intra-row spacing are the most important factors that have impact on the seed yield. Likewise, sowing dates, methods and rates are important for the main crops in Gezira Scheme. Radiation and light quality may affect Sunflower development from the flowering until seed formation stage. However, during these stages the maximum temperature was high from the end of January till the end of February, but a significant effect of air temperature on plant development was not observed (Figure 8.10). Intra-row spacing may affect the utilization of light, water and nutrients. In most cases, these factors are confounded by their effect on growth development. The significance of radiation and light quality as a factor accelerating reproductive development of Sunflower under field conditions is unknown. The magnitude of the effect of these factors was observed in our experiments, mostly in the narrow intra-row plant spacing due to competition between the plants. However, yield and yield components are affected as a result of wider row spacing. In this study the intra-row plant spacing of 30 cm (S1) represented the higher plant density of 4.2 plants/m². While the intra-row plant spacing of 40 cm (S2) represented the lower plant density of 3.1 plants/m². Seed yield and seed weight increased significantly by increasing the intra-row plant spacing from 30 cm to 40 cm, which means decrease in plant density as reported as well by Steer *et al.* (1986).

The only yield component that was not significantly affected was the stem diameter, which was the same in the two intra-row plant spacings for the two growing seasons. The response of oil content to different irrigation treatments and intra-row plant spacing showed no clear pattern. The seed number was inversely proportional with the increase in plant density. Plant density had a significant effect on the growth characteristics of Sunflower. Moreover, both 100-seeds weight and the number of seeds per head decreased significantly with increasing plant density. Sedghi *et al.* (2008) reported that crop parameters, such as 100-seeds weight and number of seeds per head, decreased when plant density increased. In this study water stress during the flowering and seed filling stages had the greatest effect on the seed yield of Sunflower as was reported as well by Erdem and Delibas (2003). The highest seed yield was obtained from the control treatment of full irrigation (weekly and 10 days irrigation intervals). Similar results were recorded by Ali (2013). The increasing severity of water stress decreased seed weight because of reduction of the seed filling stage.

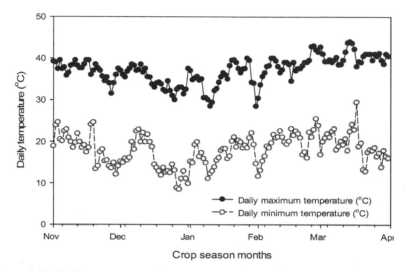

Figure 8.10. Daily maximum and minimum temperature (°C) for winter 2012 season Days to 50% flowering

Irrigation intervals showed a significant effect on the days to 50% flowering in both seasons. Under an irrigation interval of 20 days 50% flowering was reached four and seven days earlier than the other treatments (Figure 8.11). The days from sowing to 50% flowering were 55, 58, 62 days under irrigation at 20, 15 and 10 days intervals. In contrast, no significant effect of intra-row plant spacing on 50% flowering was observed during this stage. However, from the field observations the days to complete the flowering and start of the seed filling stage were between 9 and 11 days.

Figure 8.11. Effect of irrigation intervals on 50% flowering

Days to maturity

Plant spacing showed no significant effect on days to maturity in all treatments, while the irrigation intervals had a clear effect on them. The plants reached physiological maturity earlier (37 day) after the initiation of flowering (DAS 82 days) when crops received limited irrigation under water stress (20 days interval); whereas plants, which received frequent irrigation at 10 days interval reached physiological maturity late (98 days).

8.4.1   Seed yield (kg/ha)

The highest seed yield was that of W1 followed by that of W2 and the lowest seed yield was in W3 irrespective of the intra-row plant spacing. Maximum yield values of respectively 3300 kg/ha and 3290 kg/ha were obtained under irrigation at 10 days interval with intra-row plant spacing of 40 cm in the first and second season. Moreover, increasing the irrigation interval to 20 days reduced seed yield by more than 43 and 54% in the first and second season respectively under 40 and 30 cm intra-row plant spacing. The lower seed yield with 30 cm intra-row plant spacing was probably due to high competition between plants (Table 8.11). Limited irrigation water applied during the growth stages can significantly increase yield even during the heading, flowering and seed filling stages (Unger, 1983). Seed yield was significantly affected by water stress during the critical growth stages as was also found by Kazemeini *et al.* (2009). However, treatments, which were stressed during the two growth stages had a yield reduction of 5% under W2 to 43% under W3 compared to the non stressed treatments. In contrast, timing of deficit irrigation had an effect on the final seed yield as was also found by Ayana (2011).

Table 8.11 shows the interaction effects of irrigation interval and intra-row plant spacing on some crop parameters. The results showed that the 40 cm (S2) intra-row plant spacing in all irrigation treatments achieved the highest seed yield and the largest head diameter. The larger head diameter was observed under the wider intra-row plant spacing of 40 cm with irrigation at 10 days interval (W1S2). This was due to a better environment created under that treatment, which resulted in a better performance of the crop. While taller plants and the smaller head diameter were recorded under closer intra-row plant spacing, this could be due to the competition of plants for nutrients, light, moisture and carbohydrates, and also due to lack of enough space for lateral growth. On the other hand, the extended irrigation interval of 20 days W3 had a significant effect on head diameter and other components as reported as well by Kakar and Soomro (2001) and Taherabadi *et al.* (2013). In contrast, intra-row plant spacing significantly affected seed yield and other yield components, this result was similar to the findings of Ahmed. M. (2010). Results in Table 8.11 show that the plant height showed a significant difference between the full irrigated and stressed treatments. Imposition of water stress during the flowering and seed filling stages caused a high reduction in seed yield compared to full watered plants.

Results of the analysis of variance did not show any significant interaction of intra-row plant spacing and irrigation treatment on 100-seeds weight and stem diameter in the second season as was also found by Mohammad (2005). Vijayalakshmi *et al.* (1975) found that the number of seeds per head and 100-seeds weight decreased exponentially as plant population increased.

Another important parameter that can affect quantity of seed yield is head diameter. The results showed that there were significant differences among irrigation intervals regarding head diameter. Irrigation intervals of 10 days and 15 days within intra-row plant spacing of 40 cm gave the bigger head diameter (23 cm) compared with the smaller one under irrigation at 20 days interval in both intra-row plant spacings. Head diameter increased with increased irrigation frequency (Farahvash et al., (2011); Mohammed, (2013). There is a linear relationship between head diameter and the amount of irrigation levels (Taha et al., 2001). The interaction effect between intra-row plant spacing and irrigation intervals significantly affected head diameter and other parameters. The full irrigation obtained larger head diameter of 8% and 16% than irrigation intervals of 15 and 20 days respectively in both seasons.

Table 8.11. Mean of the interaction effect of irrigation treatments and intra-row plant spacing on some crop parameters and total seed yield (kg/ha) for two growing seasons

Irrigation /plant spacing treatment	Plant height (cm)	Head diameter (cm)	Stem diameter (cm)	Number of filled seeds	Total yield (kg/ha)	Weight of 100 seeds (g)
			2012/2013			
W1S1	160a	17ab	1.2a	1020a	3230a	5.1a
W1S2	161a	21a	1.4a	1180a	3300a	5.2a
W2S1	150ab	18ab	1.2a	1060a	3120b	4.8a
W2S2	152ab	21a	1.6a	1090a	2910b	5.1a
W3S1	126b	15b	1.2a	663b	1890d	4.6a
W3S2	131b	17ab	1.4a	928ab	1890d	5.1a
Mean	146	18	1.3	990	2720	5.0
SE±	3.79	0.71	0.15	62.4	165	0.39
CV%	4.49	6.62	20.1	11.2	10.5	7.53
			2013/2014			
W1S1	164a	20b	1.2a	1180a	3210a	5.8a
W1S2	157ab	23a	1.3a	1270a	3290a	6.0a
W2S1	153ab	20b	1.2a	1120ab	2860ab	5.3a
W2S2	151ab	20b	1.2a	1220a	3050ab	5.2a
W3S1	144b	17b	1.0a	1000b	1830b	5.1a
W3S2	138b	19b	1.0a	960b	1590c	5.2a
Mean	151.5	20.0	1.15	1120	2550	5.4
SE±	5.80	0.47	0.05	40.6	96.6	0.12
CV%	4.71	4.12	9.70	7.01	6.31	3.82

Means followed by the same letters are not significantly different at (P ≤ 0.05) level according to Tukey's HSD.

8.4.2 Number of filled seeds per head

The number of filled seeds per head is an important parameter to evaluate the seed yield of Sunflower, because it is considered as a direct factor. Results showed significant differences in the number of filled seeds per head among irrigation treatments and intra-row plant spacing (Table 8.12). The irrigation interval of 10 days and intra-row plant spacing of 40 cm had the highest number of filled seeds. Overall full irrigation increased the number of seeds per head by 11% and 24% compared to the 15 and 20 days intervals. Moreover, results concluded that the number of filled seeds increased linearly with each increase in irrigation frequency as was also found by Ghani et al. (2000). The number of filled seeds increased at appropriate densities, which finally led to higher seed yield.

Table 8.12. Effect of irrigation interval, intra-row plant spacing and their interactions on seeds number and total seed yield (kg/ha) for two winter seasons

	Number of filled seeds						Total seed yield (kg/ha)					
Treatment	W1	W2	W3	Means	SE±	CV%	W1	W2	W3	Mean	SE±	CV%
						2012/2013						
S1	955b	1000b	663b	874	36.0	11.21	3230b	3120a	1890a	2750	59.5	6.47
S2	1150a	1080a	865a	1030			3300a	2910b	1890b	2700		
Mean	1050	1040	764				3270	3010	1888.5			
SE	24.8						107					
CV%	6.30						9.52					
						2013/2014						
S1	1180a	1120b	1001a	1100	26.3	7.01	3210b	2860b	1830a	2670	56.4	6.36
S2	1270b	1240a	960b	1150			3290a	3050a	1590b	2710		
Mean	1220	1180	980				3250	2950	1711			
SE±	38.9						143					
CV%	8.47						13.2					

Means followed by the same letters within the same column are not significantly different at (P ≤ 0.05) level according to Tukey's HSD

Decreased soil moisture availability in the root zone by increasing the irrigation interval to 20 days resulted in the highest number of unfilled seeds. This confirmed the findings of Mehrpouyan *et al.* (2010). The results revealed that the effect of intra-row plant spacing was significant on yield and yield components of Sunflower. However, all vegetative and yield parameters were significantly affected by water storage in the soil profile due to omitting irrigation during the sensitive development stage as was found as well by Unger (1983).

8.4.3   Oil and protein content (%)

Table 8.13 displays the means of the oil and protein content obtained from different treatments. No significant differences were detected between the three irrigation treatments, the two intra-row plant spacings and their interactions. The results did not show any significant effect of irrigation levels on oil content of Sunflower while an increase in protein content was observed under mild water deficit of the 15 days interval, this is similar to the findings of Razi and Assad (1999), Kaya and Kolsarici (2011) and of Alahdadi and Oraki (20130. The results indicate that water stress significantly decreased the yield and its components but that the oil content did not change significantly. Mula Ahmed *et al.* (2007) also found that water stress had an insignificant effect on oil content. The highest oil percentage (40%) was recorded when Sunflower was irrigated every 10 days under both intra-row plant spacings in the two seasons, while the lowest oil content (38%) was recorded under the irrigation interval of 20 days. Figure 8.12 shows the results of the two seasons in oil content (%), which clearly illustrates that there were no significant differences between the two seasons.

Table 8.13. Effect of irrigation interval, intra-row plant spacing and their interactions on oil seed yield and protein content (%) for two winter seasons

Treatment	Oil seed yield (%)					Protein content (%)				
	S1	S2	Mean	SE±	CV%	S1	S2	Means	SE±	CV%
					2011/2012					
W1	40a	40a	40	0.84	5.28	14a	14a	14	0.48	7.80
W2	38a	40a	39			13a	14a	14		
W3	38a	38a	38			13a	13a	14		
Mean	38	39				13.3	13.6			
SE±	0.73					0.57				
CV%	5.65					11.3				
					2012/2013					
W1	40a	40a	40	0.82	5.20	14a	15a	15	0.35	7.30
W2	38a	38a	38			14a	15a	15		
W3	39a	38a	38			14a	14a	14		
Mean	38.9	38.8				14.1	15			
SE±	0.70					0.40				
CV%	5.42					6.83				

Means followed by the same letters within the same column are not significantly different at (P ≤ 0.05) level according to Tukey's HSD

Similarly, an increase in irrigation frequency resulted in a maximum growth rate and thus in dry matter. Water stress reduced the dry matter production under the stressed treatments by 16 - 22% in the development stage, 20 - 27% in the mid season stage and by 56 - 49% in the late season stage (Table 8.14). The most significant increase in dry matter was observed between 45 and 65 days after planting. Under full irrigation the growth rate of the plants increased gradually from the initial stage to the mid season stage. However, thereafter the growth rate decreased rapidly, indicating a reduction in vegetative growth. In addition, water deficit during the crop establishment stage substantially reduced the stomatal conductance and leaf photosynthesis. This may have led to reduced water loss from the plants. Water stress can close the stomata, which will decrease CO_2 intake and dry matter production. However, shoot development in plant height and dry matter accumulation was reduced. Throughout the crop development stages root dry weight and dry matter weight remained greater in plants at low (S2) than at high (S1) plant population densities.

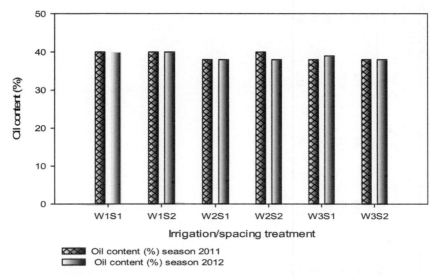

Figure 8.12. Oil content (%) for two growing seasons 2012/2013

8.4.4   Total applied water (TAW)

The field trials during the winter seasons of 2011/2012 and 2012/2013 were replicated three times in a split plot design at 18 plots. Water was applied separately to each plot and was measured during the whole growing season. Average water applied per irrigation was calculated for each treatment in m^3/ha (as mentioned in the previous Chapter). Seasonal water applied and the reduction in seed yield under different irrigation treatments and intra-row plant spacing are presented in Table 8.15. The different irrigation treatments started after the third irrigation when the plants completed the establishment stage. For treatment W3 the reduction in seed yield was 43 to 48% compared to the full irrigation treatment, while the reduction in seed yield for treatment W2 (moderate stress) was between 6 and 13%, owing to sufficient water availability during the sensitive stages. Treatments, which were stressed during the critical growth stages had a larger seed yield reduction compared to full irrigation treatments. The percentage of water saving under different irrigation treatments, that had no significantly difference in seed yield compared to the full irrigation treatment, was 31 to 36% in the first and second season respectively.

Table 8.14. Growth rate and crop development during different crop stages under different irrigation treatments (2013)

Irrigation /plant spacing treatment	Initial		Development		Mid season (1)		Mid season (2)		Late season	
	Dry matter (g)	Root (g)	Dry matter (g)	Root (g)	Dry matter (g)	Root (g)	Dry matter (g)	Root (g)	Dry matter (g)	Root (g)
W1S1	97	2.4	155	22	240	36	446	32	380	25
W1S2	107	2.8	208	29	263	48	514	51	405	30
W2S1	99	2.7	174	24	255	33	427	40	355	36
W2S2	105	3.0	193	27	294	52	573	59	403	43
W3S1	92	3.2	163	21	195	41	371	36	230	32
W3S2	98	2.9	174	26	220	40	412	43	198	30

Table 8.15. Number of irrigations and amount of water applied for each treatment and yield reduction for the winter seasons 2012/2013 and 2013/2014

Irrigation/ plant spacing treatment	Irrigation water applied (m³/ha)	Water saving (%)	Number of irrigations	Average per irrigation	Seed yield (kg/ha)	Seed yield reduction (%)
			2012/2013			
W1S1	9120	-	9	1010	3230	3
W1S2	9220	-	9	1030	3300	0
W2S1	6330	31	6	1060	3120	6
W2S2	6380	31	6	1060	2910	12
W3S1	6300	32	6	1080	1890	43
W3S2	6470	30	6	1080	1890	43
			2013/2014			
W1S1	9780	-	10	978	3210	2
W1S2	9730	-	10	973	3290	0
W2S1	6600	36	6	1100	2860	13
W2S2	6500	36	6	1080	3050	9
W3S1	5730	41	5	1146	1830	43
W3S2	5670	42	5	1134	1590	48

8.5 Comparison between summer and winter seasons

Figure 8.13 shows that there were no significant differences in total water applied to two varieties, which indicates that Bohooth-1 used a more or less similar amount of water as Hysun 33. Sunflower seed yields obtained under the different irrigation treatments were comparable among the two seasons. For comparison of the summer and winter season, it is of importance how much higher seed yield can be achieved with less water. Thus, Sunflower was sown with the same irrigation treatments in the 2012 and 2013 summer seasons (Figure 8.14), and in the winter seasons of 2013 and 2014.

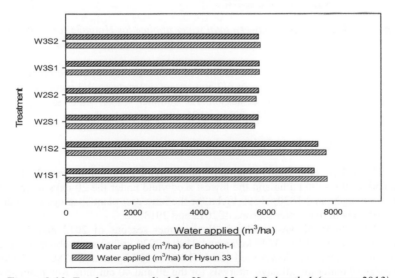

Figure 8.13. Total water applied for Hysun33 and Bohooth-1 (summer 2013)

The results show that in 2013 the total amount of irrigation water at 10 days interval and intra-row plant spacing of 30 cm was 9780 m³/ha for the winter and 7820 m³/ha for the summer season. The corresponding seed yield values were 3210 and 3100 kg/ha respectively. This means that the total amount of irrigation water for the winter season compared to the yield was considered as excess water for irrigating Sunflower. This could be attributed to differences in climatic conditions, such as ETo which was higher in the winter season than in the summer season and the contribution of rainfall during the summer season. However, it is better for saving water to irrigate other crops in the area and to cultivate Sunflower in the summer season. While for the 15 days interval the amount of water applied was 6600 and 5660 m³/ha and the corresponding seed yield values were 2860-2500 kg/ha for the winter and summer seasons respectively. Hence the lowest amount of water applied under water deficit at 20 days interval was 5730 and 5800 m³/ha for the winter and summer seasons respectively. The corresponding seed yields obtained were 1830 - 1780 kg/ha respectively for the intra-row plant spacing of 30 cm (Table 8.16).

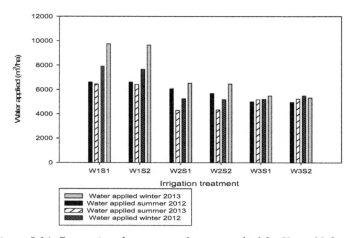

Figure 8.14. Comparison between total water applied for Hysun33 for winter and summer seasons

Table 8.16 shows that the results of water saving when compared to full irrigation W1 with mild water stress W2 and severe water stress W3 were 31 - 36% and 30 - 42% respectively for winter season. For summer season water saving from the corresponding treatments was 14 - 28% and 25 - 28% respectively.

8.6   Concluding remarks

The highest seed yields obtained from the 10 days interval were 3230 - 3300 kg/ha and 3210 - 3290 kg/ha and the lowest seed yield under the 20 days interval were 1890 - 1890 kg/ha and 1830 - 1590 kg/ha for the intra-row plant spacings of 30 and 40 cm respectively in the winter seasons of 2012 and 2013.

The yield trend was similar in the summer seasons of 2012 and 2013. The highest seed yields (3180 - 3240 kg/ha and 3100 - 2920 kg/ha) were obtained from the 10 days interval, whereas lower seed yields (1840 - 1820 kg/ha and 1780 - 1870 kg/ha) were obtained under the 20 days intervals for the 30 and 40 cm intra-row plant spacing treatments.

Table 8.16. Comparison of total water applied for Sunflower (Hysun 33) between winter and summer seasons

Irrigation /plant spacing treatment	Water applied (m³/ha)	Number of Irrigation	Average irrigation (m³/ha)	Water saving (%)	Water applied (m³/ha)	Number of irrigation	Average irrigation (m³/ha)	Water saving (%)
	Winter season 2012/2013				Winter season 2013/2014			
W1S1	9120	9	1010	-	9780	10	978	-
W1S2	9220	9	1030	-	9730	10	973	-
W2S1	6330	6	1060	31	6600	6	1100	33
W2S2	6380	6	1060	31	6500	6	1080	36
W3S1	6300	6	1080	32	5730	5	1146	41
W3S2	6470	6	1080	30	5670	5	1134	42
	Summer season 2012				Summer season 2013			
W1S1	7730	7	1104	-	7820	7	1117	-
W1S2	7740	7	1106	-	7790	7	1112	-
W2S1	6640	6	1107	14	5660	5	1130	28
W2S2	6620	6	1103	14	5700	5	1140	27
W3S1	5660	5	1132	27	5800	4	1450	26
W3S2	5610	5	1122	28	5830	4	1460	25

Compared to the maximum yield, 15 to 29% lower yields were obtained under treatments that were subject to mild water deficit (15 days interval) after crop establishment till physiological maturity. Whereas, a reduction between 40 and 44% in seed yields was obtained when the plants were subject to severe water stress (20 days interval) during the whole growing season for Hysun-33.

For Bohooth-1, the highest seed yields obtained from the 10 days interval were 3200 - 3080 kg/ha and 3180 - 2940 kg/ha for the intra-row plant spacing of 30 and 40 cm for the first and second seasons respectively.

Higher oil content was obtained from weekly and 10 days intervals. However, the highest and lowest values of percentage oil content were recorded at 40 and 38% respectively under different irrigation intervals of 10, 15 and 20 days.

The water saving under the irrigation interval of 15 days, compared to full irrigation, ranged between 31 and 36% and for the 20 days interval it ranged between 30 to 42% in the first and second winter season respectively. While the same treatments saved water by 14 - 28% and by 25 - 28% under rainy conditions (summer season).

Treatments under severe water stress saved a higher amount of water by 42%, but the yield reduction was also high, between 38 and 44%, compared to the full irrigation treatment.

9 Water productivity (WP) for Sunflower

9.1 Introduction

While agriculture is a major user of water it is facing a challenge to produce more food with less water. This requires an increase in water productivity (WP), which is an indicator of production level per unit volume of water. In the agricultural sector there is an urgent need to use available water resources efficiently and enhancing water productivity (WP). Water productivity is important for understanding water-plant relationships. Water productivity can be enhanced in many ways of water conservation techniques, such as partial irrigation, drip irrigation and deficit irrigation. Generally, these techniques appear to increase the water productivity, but with a significant yield reduction. Previous studies have been conducted to study the water productivity for different crops under various water supply levels (Zhang, 2003); Ali *et al.*, 2007; Vazifedoust *et al.*, 2008). In this study two methods have been applied to estimate the water productivity, field experiments and application of a crop growth model. In the field experiments the water productivity (WP) was estimated by measuring the water applied and crop yield (Chapters 9 and 10). Also by the crop growth model evapotranspiration (ETc) and crop yield can be simulated accurately; hence the water productivity can be estimated.

9.2 The concept of water productivity

9.2.1 The term water productivity (WP)

There are many definitions of the term water productivity, which may express a physical ratio between yields and water use (Kijne *et al.*, 2003a), or between the value of the product and water use. Therefore, it is essential to define the concept of water productivity in different ways. Increasing water productivity may be the best way to achieve efficient water use. Pereira *et al.* (2002) define water productivity as the ratio between the actual yield achieved and the total water use. Another definition for water productivity is the ratio of the amount water used to achieve a given output. Water productivity is therefore referring to crop yield per unit rainfall, total biomass per unit irrigation water (kg biomass per m^3 or kg per mm). It can be defined at different spatial and temporal (daily, weekly, seasonal and yearly) scales.

9.2.2 Study area and experiment lay out

As presented in Chapter 7 the experimental site and description were designed to investigate the effect of different irrigation treatments on yield and water productivity of Sunflower. Thus, in this Chapter water productivity and economic water productivity results will be detailed here.

Water productivity is a useful indicator for quantifying the impact of irrigation scheduling decisions with regard to water management. Water productivity (WP) is a quantitative term used to define the relationship between the crop produced and the amount of water involved in the crop production, and is defined as the amount of crop produced per volume of water applied (Molden *et al.*, 2010). The unit of WP is kg/m^3. WP is useful for analysing the potential increase in crop yield that may result from increased water availability.

Table 9.1 demonstrates the water productivity for Sunflower under different irrigation treatments. Water productivity was higher for weekly irrigation in both seasons, but in the third season (0.36 kg/m^3), it was higher under the mild stressed treatment (15 days interval). In the first season, higher water productivity of 0.26 and 0.25 kg/m^3 was obtained from irrigation at 10 days interval before the flowering and during the seed filling stage. Water productivity was low when Sunflower received irrigation at 20 days interval during the two respective stages. The reduction was 20 and 25% in the first and second season respectively. Therefore, each additional m^3 of water yielded 0.26 kg of seed, whereas the output of the other irrigation applications was much lower for each additional unit of water, 0.21 kg/m^3 for the 20 days interval during the two stages in the first season respectively, whereas there was a higher water productivity in the third season. In the third season under irrigation treatments of F3 and S3 a lower water productivity of respectively 0.26 and 0.27 kg/m^3 was obtained. Higher water productivity was obtained under moderate water stress in the third season, this confirmed that water productivity can be improved with less water as reported as well by Ali and Talukder (2008).

Table 9.1. Water productivity (WP) for different irrigation treatments for three growing seasons

Irrigation treatment	WP (kg/m^3)	WP (kg/m^3)	WP (kg/m^3)
	2011	2012	2013
Every week (W)	0.26	0.34	0.32
10 days after 50% flowering (F1)	0.25	0.31	0.35
15 days after 50% flowering (F2)	0.23	0.29	0.36
20 days after 50% flowering (F3)	0.21	0.26	0.25
10 days after seed filling stage (S1)	0.26	0.30	0.34
15 days after seed filling stage (S2)	0.22	0.28	0.33
20 days after seed filling stage (S3)	0.21	0.27	0.32

Water productivity can be improved as a result from enhancing the provision and management of the other input factors of crop production. Increased water productivity of irrigated crops is possible through proper irrigation scheduling by providing water that needs to maintain the evapotranspiration not less the crop water requirement and providing irrigation at critical stages. Figure 9.1 demonstrates water productivity for three growing seasons, which showed that the higher WP was obtained in the third season while the lower one was obtained in the first season.

9.3   Economic water productivity (EWP) US$/m^3

Agriculture plays a significant role in the economy and the progress of Sudan, among others, while about 84% of the workforce is employed in agriculture. Since independence great efforts have been made with respect to agricultural development. There are many agricultural products, which are exported (Arabic Gum and Cotton) and in return valuable foreign exchange is earned, which increases the economic system of the country. Hence, it is important to encourage and further develop the agricultural sector. In this context the calculation of economic water productivity and economic analyses were done to evaluate Sunflower seed yield under different irrigation levels and under deficit irrigation.

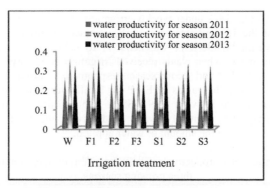

Figure 9.1. Water productivity for Winter Sunflower for three growing seasons (2011, 2012 and 2013)

9.3.1 Economic water productivity (EWP) for experiment One

The ratio between seed yield, price and water applied is denoted as economic water productivity (EWP). Water productivity expressed in physical terms only does not reflect the economic impact of water (Rodrigues and Pereira, 2009). Alternatively, crop yield may be transformed into a monetary unit (US$/ha) (Playán, 2006). Potential crop yield can be estimated from the information on regional and local weather data. Economic water productivity can be calculated by using field experiment collected data on yield values, production costs, water costs, commodity prices and irrigation performance. Water is a limiting factor in some areas; to maximize EWP may be economically more profitable for the farmers than maximizing yield (English and Raja, 1996).

In this study the economic water productivity for Sunflower was calculated by using extensive data collected at the experimental field such as the total seed yield obtained under different irrigation treatments and from three experiments, which were conducted under different climatic conditions. Also data for the seed price under local market conditions and total water applied during the whole seasons. In general the price of Sunflower seed is very low because Groundnut seed is preferred for oil extraction in some areas. The local market price was estimated with an average of 35 kg per sack.

Since the water fees are very low in Gezira Scheme, due to the high subsidies from the Government, only the increase of water charges might be reducing the amount of water use for irrigation and substantially decrease farmer's income. Previous studies indicated that the economic water productivity for the main crops in Gezira Scheme is low, substantially due to inefficiencies in the irrigation system as a result of over-irrigation (FAO, 2008). The following equation was used to calculate the economic water productivity.

EWP (US$/m^3) = Gross income (US$/ha) /Irrigation water applied (m^3/ha)   (9.1)

Table 9.2 and Figure 9.2 show the economic water productivity of Sunflower obtained from the three growing seasons. The results indicate that the EWP was generally very low. Highest economic water productivity of 0.30 US$/m^3 was obtained from weekly and 10 days irrigation in the first season. While the lowest values of EWP of 0.22 and 0.19 US$/m^3 were obtained from prolonged irrigation at 20 days interval after flowering in the first and second season respectively. Moreover, EWP was noticed to be higher by 14 and 24% under weekly irrigation (full watering) than under mild water stress under the 15 days interval and the stressed treatment under the 20 days interval after flowering. While EWP was about 17 to 21% higher under full irrigation than under the treatments when plants received irrigation at 15 days interval after the seed filling stage in the first and second season respectively. Various levels of irrigation water supplies influenced the values of EWP for Sunflower in the scheme, leading to average values that varied between 0.19 US$/m^3 for the higher water stress to 0.30 US$/m^3 for full irrigation. Generally, water productivity and EWP follow the same trend and they are related to the amount of water applied.

Table 9.2. Economic water productivity (EWP) for different irrigation treatments for the three growing seasons

Irrigation treatment	EWP (US$/m^3)	EWP (US$/m^3)	EWP (US$/m^3)
	2011	2012	2013
Every week (W)	0.30	0.29	0.25
10 days after 50% flowering (F1)	0.29	0.26	0.26
15 days after 50% flowering (F2)	0.27	0.25	0.27
20 days after50% flowering (F3)	0.24	0.22	0.19
10 days after seed filling stage (S1)	0.30	0.25	0.26
15 days after seed filling stage (S2)	0.25	0.23	0.25
20 days after seed filling stage (S3)	0.24	0.23	0.24

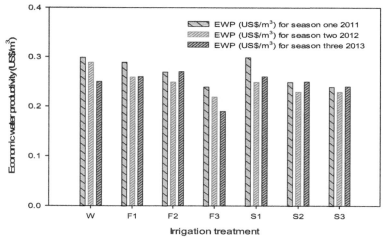

Figure 9.3. Economic water productivity for Sunflower (Hysun33) for three seasons (2011, 2012 and 2013)

Calculation of the economic analysis

Although water is available in Gezira Scheme, water supply during the peak of the season becomes limited. In this study area the land is abundant. Under this situation the economic analysis for different irrigation strategies with considering the cost of water and yield was performed. The net income, which could be achieved from the different irrigation levels, was determined by using the following equations:

Total gross return (US\$/ha) = Yield (kg/ha) × average product price (US\$/kg)   (9.2)

Total cost (US\$/ha) = variable cost (US\$/ha) + Fixed cost (US\$/ha)   (9.3)

Irrigation cost is considered here as a variable cost, which was calculated from irrigation number, irrigation labourers and fuel (pump) for irrigation at the peak of shortage in water supply. The fixed costs are the production input costs during the growing season, which include seeds cost, irrigation cost, fertilizer, ploughing, weeding, cultivation, harvesting and the cost of other inputs. Equations 9.4 and 9.5 were used to calculate the net benefit or net income and benefit-cost ratio under different irrigation levels to assess to which extent the application of water deficit would have resulted in more profit to the farmers without any significant reduction in marketable yield.

Net benefit (US\$/ha) = Gross return (US\$/ha) - Total cost (US\$/ha)   (9.4)

Benefit - cost ratio = Gross income/Total cost   (9.5)

Results of the economic analysis of Sunflower production under different irrigation intervals are presented in Tables 9.3 and 9.4. To assess economically optimum crop water application that maximizes returns of irrigation water under different irrigation regimes the average of the seed yield was obtained from each plot in kg/ha and the average market price per kilogram for each growing season was estimated at US\$ 1.19 /kg in the first season and at US\$ 0.83/kg in the second and third season. The net return (net benefit) was computed by subtracting the total cost, which included fixed cost and production expenses of cultivation from the gross income of the production. The inputs in all treatments were the same during all the growing seasons. The only difference has been in the cost of the water labourers per irrigation, which depended on the number of irrigations in each treatment. According to the results, the maximum net income of 1270 US\$/ha was obtained from the full irrigation treatment (W), whereas the lower net incomes of 288 and 327 US\$/ha were obtained from the stressed treatments F3 and S3 respectively, owing to the low seed yield. The net income decreased to 18% (0.21 US\$/ha) when the crop was irrigated at 10 days interval after the flowering stage and total costs decreased by 3% from the full irrigation treatment.

The estimation of the market prices depends on the rate of the dollar during the harvesting time. Although there were fluctuations in the market prices the maximum net income of 799 US\$/ha was obtained from the full irrigation treatment with a high benefit-cost (B/C) ratio of 1.4. The lowest net income of 48 US\$/ha with a B/C ratio of 1.0 was obtained under the water stress treatment (F3) in the second season (2012). In contrast, the maximum ratio of net income of 1.5 was obtained from the F1 and S1 irrigation treatments, while the full irrigation treatment gave less net income compared to F1 and S1, owing to the high inputs with low outputs in the third season (Table 9.4). The average benefit-cost ratio was 1.5, indicating that Sunflower is a profitable crop under the current situation. Full irrigation increased farmers' incomes by 7% compared with deficit irrigation, which decreased the B/C ratio by 13 and 20%. The results show that the traditional method of the farmers in Gezira Scheme is to irrigate their farms with exceeding water to avoid crop failure. Figure 9.3 illustrates the high production cost in the first season compared to the second season. The net return or benefit provides low values for the second year, although in the last year it reached higher values due to an increase of local market prices. Whereas seed yield and net return was similar in both seasons. This could be due to the fluctuation in market prices.

Table 9.3. Economic analysis for different irrigation treatments for the winter season 2011

Irrigation treatment	Yield (kg/ha)	Market price (US$/kg)	Gross return (US$/ha)	Variable cost (US$/ha)	Fixed cost (US$/ha)	Total cost (US$/ha)	Net return (US$/ha)	B/C ratio
W	3130	1.19	3730	1190	1280	2460	1270	1.5
F1	2860	1.19	3410	1100	1280	2370	1040	1.4
F2	2410	1.19	2880	1100	1280	2370	502	1.2
F3	2150	1.19	2570	1000	1280	2280	288	1.1
S1	2800	1.19	3350	1100	1280	2370	975	1.4
S2	2350	1.19	2810	1100	1280	2370	435	1.2
S3	2190	1.19	2610	1000	1280	2280	327	1.1

Table 9.4. Economic analysis for different irrigation treatments for the two winter seasons 2012 and 2013

Irrigation treatment	Season 2012					Season 2013				
	Yield (kg/ha)	Gross return (US$/ha)	Total cost (US$/ha)	Net return (US$/ha)	B/C ratio	Yield (kg/ha)	Gross return (US$/ha)	Total cost (US$/ha)	Net return (US$/ha)	B/C ratio
W	3140	2590	1790	799	1.4	3100	2560	1790	770	1.4
F1	2880	2370	1710	667	1.4	3070	2540	1650	888	1.5
F2	2510	2070	1710	367	1.2	2670	2200	1580	627	1.4
F3	2130	1750	1710	48	1.0	2260	1860	1580	285	1.2
S1	2840	2350	1790	559	1.3	2970	2450	1650	801	1.5
S2	2350	1940	1710	234	1.1	2490	2050	1580	475	1.3
S3	2270	1870	1710	164	1.1	2400	1980	1580	408	1.3

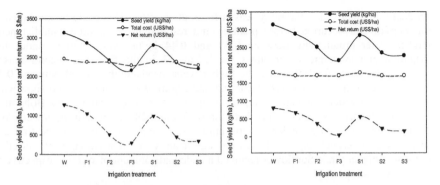

Figure 9.3. Relationship between seed yield, total cost and net return for Sunflower (Hysun33) for the two winter seasons 2011 and 2012

9.4   Concluding remarks on experiment One

Water productivity in terms of water applied, economic water productivity (EWP) and economic returns were computed for irrigated Sunflower based on field experimental data. Water productivity was low when severe water stress occurred after flowering and seed filling. The reduction was 20 and 25% in the first and second season respectively. Water productivity was higher when the plants were irrigated every week in both seasons. However, in the second season (0.34 kg/m^3) it was higher due to a high yield and a lower amount of water used than in the first season (0.26 kg/m^3) also at 10 days interval during the flowering and seed filling stages. Generally, the highest water productivities of 0.25 - 0.26 kg/m^3, 0.31 - 0.30 kg/m^3 and 0.34 - 0.35 kg/m^3 were achieved under irrigation at 10 days interval after the flowering and seed filling stages from the three seasons respectively. The most sensitive stages of water stress are the flowering and seed filling stages. The lowest water productivities of respectively 0.26 and 0.27 kg/m^3 were obtained under the irrigation interval of 20 days after the flowering and seed filling stages in the second season.

Water productivity values are influenced by the water management system. In this study the irrigation methods were surface and furrow irrigation, which led to lower water application efficiency in Gezira Scheme. Maximum profit of 1270 US$/ha was obtained under the full irrigation treatment in the first season, whereas the minimum net incomes of respectively 288 and 327 US$ /ha were obtained from the water stress treatments after flowering (F3) and seed filling (S3).

The maximum net income of 799 US$/ha was obtained from the full irrigation treatment with a benefit-cost ratio of 1.4 and the lowest net income of 48 US$/ha with a benefit-cost ratio of 1.0 was obtained under the stressed treatment (F3) in the second season 2012. While under the 10 days interval after flowering and seed filling stages the maximum benefit-cost ratio of 1.5 was obtained.

9.4.1   Water productivity (WP) for experiment Two (summer season)

Water productivity can be increased by increasing yield achieved with less water per unit area. In addition, in order to produce more crops with less water, water management strategies and practices need to be considered. Figure 9.4 illustrates that water productivity was higher in the summer season compared to the winter season, although there was insufficient rainfall and supplemental irrigation was applied. Water productivity improved in the rainy season for both varieties. This could be attributed to the rainfall that occurred before planting and resulted in better crop establishment. Higher water productivity between 0.38 and 0.44 kg/m^3 was obtained under full irrigation and moderate water stress for Hysun 33 in the first and second season respectively, while for Bohooth-1 the higher water productivity ranged between 0.37 and 0.41 kg/m^3 in both seasons (Table 9.5 and Figure 9.4). There were no significant differences in water productivity among both varieties. The local variety Bohooth-1 obtained a higher EWP that ranged between 0.25 and 0.28 US$/m^3 and the lower EWP of 0.19 and 0.22 US$/m^3 under full irrigation and severe water stress in the first and second season respectively. EWP was not significantly different among both varieties, but it was higher when Sunflower received an adequate amount of water irrespective to intra-row plant spacing.

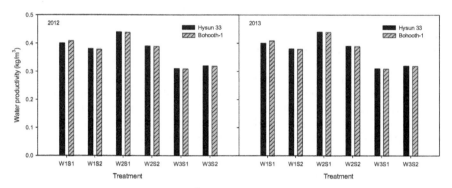

Figure 9.4. Water productivity (kg/m^3) of two Sunflower hybrids for summer 2012 and 2013

Table 9.5. Water productivity and economic water productivity for two hybrids for two summer seasons of 2012 and 2013

Irrigation /plant spacing treatment	EWP (US$/m^3)	WP (kg/m^3)	EWP (US$/m^3)	WP (kg/m^3)
Hysun 33	*2012*		*2013*	
W1S1	0.25	0.41	0.27	0.40
W1S2	0.25	0.42	0.25	0.38
W2S1	0.24	0.41	0.30	0.44
W2S2	0.21	0.34	0.26	0.39
W3S1	0.19	0.32	0.21	0.31
W3S2	0.19	0.32	0.22	0.32
Bohooth-1				
W1S1	0.25	0.41	0.28	0.41
W1S2	0.25	0.41	0.26	0.39
W2S1	0.24	0.39	0.25	0.37
W2S2	0.20	0.33	0.26	0.38
W3S1	0.19	0.32	0.22	0.33
W3S2	0.19	0.33	0.22	0.33

9.4.2   Water and economic water productivity for experiment Three (winter season)

Local farming systems, opportunities to improve Sunflower seed yield are mostly associated with management decisions made at planting. Water productivity is a key to evaluate deficit irrigation strategies. The following equation was used to calculate the water productivity:

WP (kg/m^3) = seed yield (kg/ha) / water applied (m^3/ha)   (9.5)

Water productivity (WP) can be achieved by using deficit irrigation strategies. In this study a higher WP between 0.36 and 0.49 kg/m^3 was obtained under mild water stress for the 15 days interval. Significant differences were detected between the different irrigation treatments and intra-row plant spacing interactions. WP recorded the highest values of 0.44 and 0.49 kg/m^3 under both intra-row plant spacings and the irrigation interval of 15 days for the first season and 0.36 kg/m^3 for the 10 and 15 days interval in the second season. It declined to its lowest level (0.28 kg/m^3) with intra-row plant spacing 40 cm and the irrigation interval of 20 days in the second season, due to the low yield obtained with high amount of water applied (Table 9.6). As a result, irrigation at 20 days interval led to severe crop suffering during its reproductive stage, which substantially lowered seed yield. An irrigation interval of 15 days W2 resulted in a higher water productivity of 0.49 kg/m^3, but did not differ much from water productivity of irrigation at 10 days interval W1. Despite increasing the irrigation interval from 10 to 15 days, no significant impact was observed in seed yield, but in water productivity it was obvious. Although the yield increased with increases in water use, water productivity decreased. Figure 9.5 shows the results of the calculation of the water productivity and economic water productivity plotted against different treatments for the two winter seasons 2012 and 2013. Higher water productivity of 0.46 and 0.36 kg/m^3 was obtained in the first and second seasons (2012 and 2013) under full irrigation and mild water stress, while the lower WP under severe water stress was 0.28 kg/m^3. The same trend was observed in the EWP. Similarly, higher WP and EWP were obtained from mild water stress of irrigation at 15 days interval. This suggests that water deficit strategies can increase WP and EWP.

Table 9.6. Water productivity and economic water productivity under different treatments for two winter seasons 2012 and 2013

Irrigation /plant spacing treatment	Irrigation water applied (m^3/ha)	WP (kg/m^3)	EWP (US$/m^3)	Irrigation water applied (m^3/ha)	WP (kg/m^3)	EWP (US$/m^3)
	2012/2013			*2013/2014*		
W1S1	9120	0.35	0.21	9780	0.33	0.21
W1S2	9220	0.36	0.21	9730	0.34	0.22
W2S1	6330	0.49	0.27	6600	0.43	0.27
W2S2	6380	0.46	0.25	6500	0.47	0.29
W3S1	6300	0.30	0.18	5730	0.32	0.20
W3S2	6470	0.29	0.17	5670	0.28	0.18

Field data were collected on yield values, production costs, water fees, crop prices and times of irrigation water applied. Irrigation water profit was mainly determined by the seasonal volume of water applied (variable costs). EWP was calculated and the results showed that under Gezira conditions EWP is low. However, the economic water productivity results are affected by the weak performance of the irrigation system (Table 9.6). For adopting deficit irrigation in the existing system is the main problem due to the unwillingness of farmers to reduce the water supply to their fields. Considering these data, it gives evidence that EWP values are low and the yield prices of Sunflower cover the production costs, particularly under full irrigation conditions (W1S1). Highest EWP of 0.27 and 0.29 US\$/m³ was obtained from irrigation at 15 days interval and the lowest EWP of 0.17 US\$/m³ was obtained from at 20 days interval in the first season (Figure 9.6).

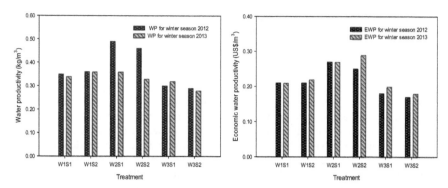

Figure 9.5 Water productivity and economic water productivity under different treatments for winter seasons 2012 and 2013

9.4.3 Economic water productivity (EWP) for experiment Two

Sunflower is considered as a lower sensitive crop to water deficit. It will be an attractive crop if the market prices become higher and the irrigation systems will be improved. The impact of water stress on yield related to economic water productivity may be positive depending upon the irrigation system, yield values and irrigation system performance (Rodrigues and Pereira, 2009). They also reported that WP and EWP are highly dependent on climate conditions. Some of the previous studies indicated that the EWP for Sunflower was low and varied from 0.11 to 0.24 €/m³.. Moreover, they assessed the economic feasibility of deficit irrigation implementing to Sunflower and water fees. Higher economic water productivity was obtained from irrigation at 10 and 15 days intervals, while severe water stress decreased economic water productivity to lower values of US\$ 0.19 and 0.22/m³. Deficit irrigation can increase the water productivity, and therefore, the economic water productivity. Figure 9.6 illustrates that irrigation at 10 and 15 days interval are considered economically feasible treatments. Achieving, higher yields with increased water productivity is only economical when increased profits are not due to increased cost of other inputs in crop yield.

9.4.4 Economic evaluation under different irrigation strategies

The financial feasibility of deficit irrigation to grow Sunflower was determined by a benefit-cost analysis with a discount rate of 18%, based on the results from the field experiments of 2012 and 2013. Net return, which is a ratio between gross income and total costs, depends mainly on production costs and water costs. In this study the water costs were low due to the high subsidies from the Sudan Gezira Board. The irrigation costs consist of the design of small ditches and irrigation labourers (water-man). However, the cost of irrigation and associated labour cost cause variation in total cost (Table 9.7).

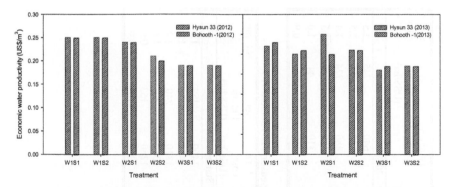

Figure 9.6. Economic water productivity (US$//m³) for two hybrids for summer 2012 and 2013

Results revealed that higher production costs under irrigation at 10 days interval of 8 - 9% compared to irrigation at 15 days interval, resulted in a higher net return of 721 US$/ha and 866 US$/ha obtained from irrigation at 15 days interval in the first and second winter seasons respectively. Reasons for the higher profits under full irrigation were mainly due to higher yields and gross income, compared to two other treatments (Table 9.7). The economic analysis showed that 32 and 42% reduction in irrigation would reduce the gross income by 41 and 43%, respectively. Therefore, the deficit-irrigation could cause substantial economic loss through decreased crop marketability.

Tables 9.8 and 9.9 show the results of the economic analysis for two hybrids for the summer seasons 2012 and 2013. In this context, results show that farmer's net return from farming Sunflower was quite low at the current low commodity prices. On the other hand, both the treatments W1 and W2 resulted in significantly more net returns compared to W3. Results also show that there was a significant drop in net return (US$ -153/ha) when the plants were exposed to water stress during all the growth stages. This could be due to the high inputs and less outputs as a result of the effect of water stress on Sunflower yield. Accordingly, the recent low market prices with high production costs would also lead to negative incomes, even with a good crop and irrigation practice management. However, the results could be positive and farmers can increase their incomes if the market prices would be higher. Results also show that the net income for Hysun 33 was 8 and 5% higher compared to Bohooth-1 when the crops were irrigated at respectively 10 and 15 days interval. Thus, it can be concluded that Hysun 33 is more profitable than Bohooth-1. Deficit treatments showed a negative net return, implying that the costs of production are higher than the gross return. The costs of production increase with increasing applied water.

Deficit irrigation treatment W3 resulted in considerable economic loss compared to W1 and W2, because the total cost of production exceeded the gross income (Tables 9.7 and 9.9). The benefit-cost ratio (B/C), which indicates gross income per cultivated land, was also influenced by irrigation schedules (Kahlown *et al.*, 2007). Irrigation at 10 days interval W1 gave a higher B/C ratio (1.5) for the 2013 winter season as well as for the 2012 summer season. Benefit-cost ratios of full irrigation treatments were higher than 1, therefore it appears that investing in Sunflower in Gezira Scheme with irrigating plants by full irrigation at 10 and/or 15 days interval is an economically viable option at current market prices. Several studies indicate that deficit irrigation applied in crop growth stages resulted in the improvement of water productivity.

Table 9.7. Economic analysis of different irrigation treatments for two winter seasons 2012 and 2013

Irrigation /plant spacing treatment	Yield (kg/ha)	Gross income (US$/ha)	Total cost (US$/ha)	Net return (US$/ha)	B/C ratio	Yield (kg/ha)	Gross income (US$/ha)	Total cost (US$/ha)	Net return (US$/ha)	B/C ratio
			2012/2013					2013/2014		
W1S1	3230	2360	1770	595	1.3	3210	2390	1640	745	1.5
W1S2	3300	2410	1770	648	1.4	3290	2440	1640	799	1.5
W2S1	3120	2280	1560	721	1.5	2860	2120	1480	724	1.4
W2S2	2910	2120	1560	563	1.4	3050	2260	1480	866	1.5
W3S1	1890	1380	1460	-75	0.9	1830	1360	1310	44	1.0
W3S2	1890	1380	1460	-77	0.9	1590	1180	1310	-130	0.9

Table 9.8. Economic analysis of different irrigation treatments for Hysun 33 for two summer seasons 2012 and 2013

Irrigation /plant spacing treatment	Yield (kg/ha)	Gross income (US$/ha)	Total cost (US$/ha)	Net Return (US$/ha)	B/C ratio	Yield (kg/ha)	Gross income (US$/ha)	Total cost (US$/ha)	Net Return (US$/ha)	B/C ratio
			Summer 2012					Summer 2013		
W1S1	3180	2370	1560	805	1.5	3100	1920	1460	459	1.3
W1S2	3240	2410	1560	846	1.5	2920	1810	1460	346	1.2
W2S1	2700	2000	1400	605	1.4	2500	1550	1310	240	1.2
W2S2	2330	1730	1400	332	1.2	2200	1360	1310	54	1.0
W3S1	1840	1370	1230	133	1.1	1780	1100	1260	-153	0.9
W3S2	1820	1350	1230	116	1.1	1870	1160	1260	-102	0.9

Table 9.9. Economic analysis of different irrigation treatments for Bohooth-1 for two summer seasons

Irrigation /plant spacing treatment	Summer 2012					Summer 2013				
	Yield (kg/ha)	Gross income (US$/ha)	Total cost (US$/ha)	Net return (US$/ha)	B/C ratio	Yield (kg/ha)	Gross income (US$/ha)	Total cost (US$/ha)	Net return (US$/ha)	B/C ratio
W1S1	3200	2380	1710	669	1.4	3080	2350	1780	573	1.3
W1S2	3180	2360	1710	655	1.4	2940	2250	1780	464	1.3
W2S1	2630	1950	1590	360	1.2	2140	1640	1630	7	1.0
W2S2	2240	1660	1590	69	1.0	2230	1700	1630	76	1.0
W3S1	1800	1340	1480	-142	0.9	1900	1450	1580	-120	0.9
W3S2	1800	1340	1480	-139	0.9	1900	1460	1580	-119	0.9

Deficit irrigation increased the benefits with more than 50% compared with the usual irrigation practice of over-irrigation by the farmers (Ali and Talukder, 2008; Geerts and Raes, 2009). These results also can be used to estimate the range of water use within which deficit irrigation would be more profitable than full irrigation.

The results of the comparison between winter and summer season for water productivity and economic water productivity of Sunflower under different irrigation regimes are presented in Table 9.10. The results show that the WP and EWP were higher in the summer season compared to the winter season (Figures 9.7 and 9.8). The differences among the various treatments are not significant. There was insufficient rainfall during this period due to the fluctuation of rainfall. Thus, supplementary irrigation was applied to the experimental plots.

Table 9.10. Comparison between water productivity and economic water productivity for Sunflower (Hysun 33) between winter and summer seasons

Irrigation /plant spacing treatment	WP (kg/m^3)	EWP (US$/m^3)	WP (kg/m^3)	EWP (US$//m^3)
Winter		*2012/2013*		*2013/2014*
W1S1	0.35	0.21	0.34	0.19
W1S2	0.36	0.21	0.36	0.20
W2S1	0.49	0.29	0.33	0.25
W2S2	0.46	0.27	0.36	0.27
W3S1	0.30	0.18	0.32	0.18
W3S2	0.29	0.17	0.28	0.16
Summer		*2012*		*2013*
W1S1	0.41	0.25	0.41	0.27
W1S2	0.42	0.25	0.38	0.25
W2S1	0.37	0.24	0.44	0.30
W2S2	0.34	0.21	0.39	0.26
W3S1	0.32	0.19	0.31	0.21
W3S2	0.32	0.19	0.32	0.22

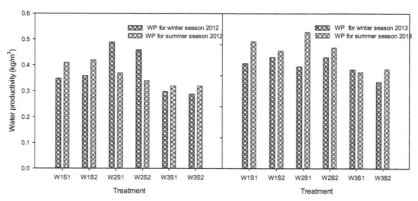

Figure 9.7. Comparison of water productivity for Hysun 33 between winter and summer seasons 2012 and 2013

The average values were 0.17 and 0.19 US$/m^3, indicating that the application of irrigation did not generate an increment in gross income. However, in the second season in 2013 Sunflower values reached their higher values of 0.27, 0.22 and 0.30 US$/m^3 for W1S1, W2S1 and W3S2 due to the increase in local market prices. Although the production cost increases in the rainy season due to continuous weeding processes, but water productivity was higher (Figure 9.7). Farmers rely on yield increase and the reduction of irrigation and cultivation labourers to increase their incomes.

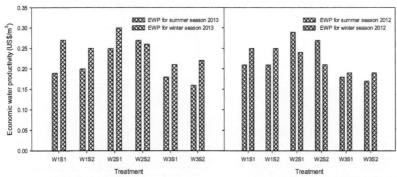

Figure 9.8. Comparison of economic water productivity for Hysun 33 between winter and summer seasons 2012 and 2013

9.5   Concluding remarks

Due to the increase in the market price of Groundnut from 0.09 to 0.18 US$/kg in 2013, many farmers preferred to cultivate Groundnuts to get a higher income. As a result the cultivated area of Sunflower has gone down from 210,000 to 126,000 ha. However, highest EWP was obtained when the market prices of Sunflower increased in 2012.

Irrigation at 15 days interval recorded the highest levels of water productivity, which ranged between 0.33 and 0.49 kg/m^3, and between 0.34 and 0.44 kg/m^3 for the winter and summer seasons of 2012 and 2013 respectively. Thus, higher water productivity under mild water stress is exhibited by the results.

The results also show that the costs increased with increasing the applied water. This is due to increased water and labour costs. Consequently, the economic analysis mainly depends on production costs, yield values; amount of irrigation water applied per hectare in which the full irrigation treatment was compared with deficit irrigation treatments.

Accordingly, the results of the economic evaluation under various irrigation strategies will provide information to policy makers for planning improvement regarding irrigation management practices.

The highest net incomes of 805 - 846 US$/ha were recorded for Hysun33, which is higher than 669 - 655 US$/ha for Bohooth-1 in one season. Also the total costs in Bohooth-1 were higher than Hysun 33. A net return was higher in Hysun 33 than in Bohooth-1 by 20% in the first summer season. Thus, it can be concluded that Hysun 33 is more financially viable than Bohooth-1 at the current market prices under Gezira Conditions.

10 Modelling of Sunflower water productivity under different irrigation intervals by using AquaCrop

10.1 AquaCrop model description

AquaCrop is FAO's water productivity simulation model resulting from the revision of the FAO Irrigation and Drainage paper No. 33 'Yield Response to Water' (Doorenbos, 1979). For over two decades this paper has been a key reference for estimating the yield response of field, vegetable and tree crops to water. To assess accurately crop yield under limited water availability, models have been required since water deficit throughout the season varies in intensity, duration and time of occurrence. AquaCrop focuses on simulating the attainable crop biomass and harvestable yield in response to water, which is the key driver for agricultural production and which becomes increasingly critical in crop production. It is a tool for:

- predicting crop production under different water management conditions (including rainfed, supplementary, deficit and full irrigation) under present and future climate change conditions;
- investigating different management strategies under present and future climate change conditions;
- soil water content, canopy cover, biomass production and final seed yield used to calibrate the AquaCrop model.

The application of the AquaCrop model was apparent and widely used in recent years to simulate yield response to water supply under different climatic conditions for a number of crops such as Cotton (Farahani *et al.*, 2009) for Sunflower (Todorovic *et al.*, 2009), for Barley, for Teff (Araya *et al.* 2010a) and for Winter wheat in Iran (Andarzian *et al.*, 2011). Steduto *et al.* (2009), Geerts and Raes (2009) and Geerts *et al.* (2010) used AquaCrop to schedule Guinoa irrigation, Araya *et al.* (2010a) to determine the optimal harvesting date of Barley and Araya *et al.* (2010b) for Teff in Ethiopia. Stricevic *et al.* (2011) reported that AquaCrop was able to simulate yield of Sunflower with a high degree of reliability. However, model performance information is relatively scarce, and it is not known yet. Moreover, it is interesting to assess the model performance in the context of productivity and yield variability under field conditions.

Doorenbos and Kassam (1979) represent an important source to determine the yield response to water:

$$1 - \frac{Y}{Yx} = Ky \frac{1-ETa}{ETp} \tag{10.1}$$

Where:

Yx, Y	= maximum and actual yield (kg/ha)
(1-Y/Yx)	= relative yield decline
ETp, ETa	= potential and actual evapotranspiration (mm/day)
(1-ETa/ETp)	= relative water stress
Ky	= proportionality factor between relative yield decline and relative reduction in evapotranspiration

AquaCrop evolves from the Ky approach by separating: i) the actual evapotranspiration (ETa) into soil evaporation (E) and crop transpiration (Tr); ii) the final yield (Y) into biomass (B) and harvest index (HI). This change in model parameters leads to the following equation at the core of the AquaCrop growth engine (Figure 10.1).

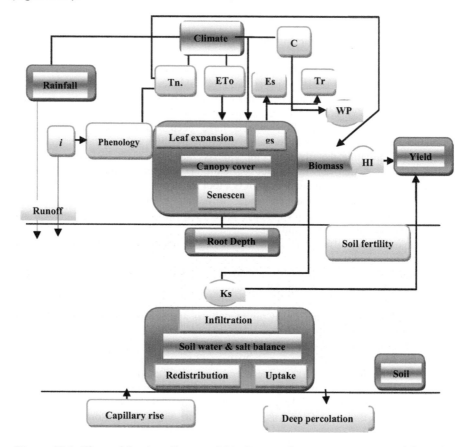

Figure 10.1. Chart of the AquaCrop model indicating the main components of the soil-plant-atmosphere Continuum (Adapted from Raes et al., 2009)

$$B = WP.\sum Tr \qquad (10.2)$$

Where:

Tr   = crop transpiration (mm)

WP   = water productivity parameter (kg) of biomass per m^3 and per mm of cumulated water transpired over the time period in which the biomass is produced (Steduto, 2007)

In previous equations a complete set of model components was inserted, including the soil with its water balance; the crop with its development, growth and yield processes and the atmosphere with its thermal regime, rainfall, evaporation demand and carbon dioxide concentration. Also, irrigation and fertilizer management are considered as they will affect the soil water balance, crop development and therefore the final yield. The model components and the relationship between model parameters are elaborated in the flow chart (Figure 10.1).

Where the crop parameters driving phenology, canopy cover, transpiration, biomass production and final yield are: I, irrigation; Tn, minimum air temperature; Tx, maximum air temperature; ETo, reference evapotranspiration; E, soil evaporation; Tr, canopy transpiration; gs, stomatal conductance; WP, water productivity; HI, harvest index; CO_2, atmospheric carbon dioxide concentration.

One of the important key features of the AquaCrop model is the simulation of green canopy instead of the leaf area index (LAI). The variation of canopy cover (CC) is expected due to the critical role in crop modelling (Duchemin *et al.*, 2008). Therefore the canopy growth coefficient (CGC) monitored at the field was used in the model instead of LAI.

The AquaCrop model provides a simple and robust method to estimate crop yield response to water with low data requirement for input parameters. Although, AquaCrop is based on complex crop physiological processes, it requires a relatively small number of outright parameters with high accuracy (Steduto *et al.*, 2009). Although other crop models have produced good crop yield simulation results, compared to them, the AquaCrop model is simpler, requires available field input data, is generally available, and is highly reliable for the simulation of biomass, canopy cover and yield under different climatic conditions.

10.2   AquaCrop model input data

To run the model input data (Figure 10.2) such as weather data, irrigation management, crop parameters - plant density, sowing date, effective rooting depth, crop canopy percentage - were collected and soil water content was measured. Canopy cover was measured by using a Plant Canopy Analyzer. Time to senescence and maturity were recorded.

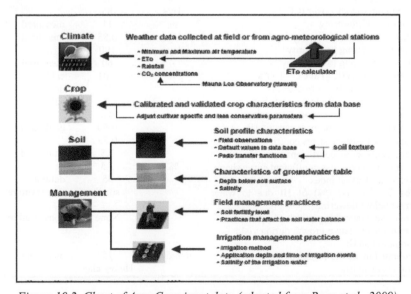

Figure 10.2. Chart of AquaCrop input data (adapted from Raes et al., 2009)

10.2.1 Climatic data

The climate of Gezira area is arid. The climatic data required for the AquaCrop model to calculate ETo were collected from Gezira Meteorological Station. The specific climatic data used to calibrate the model were; maximum temperature (°C), minimum temperature (°C), relative humidity (%), sunshine (hr) and wind speed at 2 m (m/s).

10.2.2 Crop phenological data

Data of canopy cover are important in the AquaCrop model, which was estimated from the initial canopy cover until the maximum canopy cover during the mid season. Plant canopy varied among each treatment. However, there was a direct relationship between canopy cover and irrigation treatments. After inputting the observed phenological dates, such as dates of maximum expansion, emergence and senescence, the crop cover expansion was estimated automatically. The parameters of the canopy growth coefficient (CGC), the canopy decline coefficient (CDC), stress indices and their curve shape factors, effect on leaf expansion and early senescence are presented in Table 10.1. The stress of soil fertility was not considered because sufficient fertilizer of urea was applied during the crop cycle (Table 10.2).

Table 10.1. Conservative and non-conservative parameters used in the AquaCrop model for Sunflower calibration

Parameter	Unit	Values	Description
Base temperature	°C	5	conservative
Upper temperature	°C	35	conservative
Plant density	Plants/m^2	4.2	
Time to emergence	days	6	measured
Time from sowing to maximum canopy	days	60	measured
Canopy decline (CDC)	days	13	
Time to senescence	%	85	
Maximum canopy cover (CC$_x$)	days	85	
Duration of flowering	m	11	measured
Maximum effective rooting depth	days	0.85	measured
Day from sowing to maturity	days	100	
Day from sowing to flowering	-	60	measured
Crop coefficient(K_{cb}) at CC =100%	g/m^2	1.01	calibrated
Water productivity (normalized)	-	17	default
Stomata stress coefficient curve shape	-	3	calibrated
Upper threshold for canopy expansion		0.3	
Lower threshold canopy expansion (P$_{lower}$)		0.60	calibrated
Canopy senescence stress coefficient (p$_{upper}$)	-	0.55	
Senescence stress coefficient curve shape		3.5	calibrated
Leaf expansion stress coefficient curve shape	%	3.5	
Reference harvest index		35	defalut
Possible increase (%) of HI due to water stress before flowering		small	conservative
Coefficient describing positive impact of restricted vegetative growth during yield formation on HI		None	conservative
Saturated hydraulic conductivity	mm/day	15	measured
Soil characteristics	-	Heavy clay	

10.2.3 Soil information

The clay soil in Gezira is known as black cotton soil. In this soil the field capacity and wilting point values are about 43 and 23% respectively. The total available water (TAW) is equal to 200 mm (Abdelhadi *et al.*, 2002). The saturated hydraulic conductivity is very low, about 0.03 - 0.06 cm/hr.

Table 10.2. Duration (days) of various growth stages during both seasons

Growth stage	Day after sowing (DAS)	Experiment 1	Experiment 2
Initial	22	14 Nov - 2 dec 2011	19 Nov - 10 Dec 2012
Development	35	3 Dec 2012 - 7 Jan 2012	11 Dec 2012 - 16 Jan 2013
Mid season	30	8 Jan - 8 Feb 2012	17 Jan - 16 Feb 2013
Late season	25	9 Feb - 6 Mar 2012	17 Feb - 13 Mar 2013

10.3 Calibration of the AquaCrop model

The field data of all the experiments that were conducted to assess the crop response to water stress of Sunflower (*Helianthus annuus L.*) were used to calibrate and validate the AquaCrop model. Within this research work there were two objectives:

•   to calibrate and validate the AquaCrop model using crop data obtained from the field experiments;

•   to analyze model performance and assess the applicability and accuracy of the AquaCrop model under different irrigation water and deficit conditions.

The AquaCrop model offers files, which contain parameters suitable for the simulation of Sunflower. Some of these parameters are not universal and need to be adjusted to local conditions. Crop development data were measured and derived from the experimental field observations, while other data were adjusted and calibrated. Some of the input data were locally calibrated over the year 2011. The crop parameters that were used to calibrate the model were presented in Table 10.1. These parameters are known to be conservative and applicable under all conditions (Hsiao *et al.*, 2009).

For calibration observation data were taken and used in the model. Sets of parameters were calibrated and the best results were obtained. Initial soil water content at time of planting was measured by the gravimetric method. The observed irrigation schedules were defined by specifying the date and depth of each irrigation. After the calibration canopy cover, water productivity and seed yield were compared between observed and simulated results. Biomass data for the growing season were not taken. The modelling of yield response to water is expected to play an increasingly important role in the optimization of water productivity (CWP) in agriculture. In crop simulation models, calibration is important to estimate the parameter values, which could be obtained from the model for different crops, varieties and climatic conditions. Comparative studies between simulated results with data obtained from the field experiments allowed to study model performance under drought and farming systems.

10.3.1 Reference evapotranspiration data

Reference evapotranspiration (ETo) was used to calibrate the AquaCrop model by using the Penman-Monteith formula. FAO's EToCalc software program (2009) was used to calculate ETo (Figure 10.3) and thereafter the obtained data were exported to the climatic file in the model. Climatic data, which have been considered as important data in the model are maximum and minimum air temperature (°C), mean relative humidity (%), sunshine (hr) and wind speed (m/s).

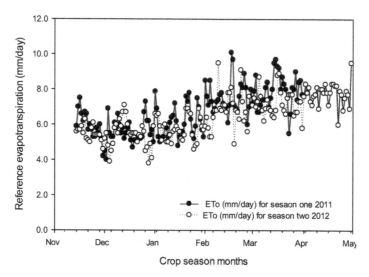

Figure 10.3. ETo computed from daily climatic data for the two winter seasons

10.3.2 Evapotranspiration data for model calibration

With the AquaCrop model the daily water balance can be calculated. The evapotranspiration (ETc) is separated into evaporation (E) and transpiration (T), which is related to canopy cover. The influence of water stress on crop development was simulated through four stress coefficients (leaf expansion, stomatal closure, canopy senescence and harvest index). The response of the crop to any water stress can occur during the growing cycle at any stage. This leads to reduction in the canopy expansion rate during the initial stage, closure of stomata during the crop development and to accelerated senescence during the late growth.

10.3.3 Irrigation management data for model calibration

The AquaCrop model contains different user-specific options for irrigation management, such as determination of the net irrigation requirement, irrigation schedule based on irrigated and rainfed management strategies. The irrigation scheduling provides different methods of application (basin, furrow, border and drip irrigation). In this study the ground was systemized for furrow irrigation so that restricted flow is not considered, nor is embankment. The date and depth of the irrigation was directly inputted into the model. In the present study the irrigation water was applied for each treatment in different quantities during the growing season. Deficit irrigation was imposed after the flowering and seed filling stages, which caused strongly and slightly impacts on seed yield and water productivity respectively. Irrigation water applied during the crop growing cycle was used to calibrate the model (Figure 10.4). Data from the experiments were calculated and exported to the irrigation file in the model.

10.4   Validation of the AquaCrop model

Before using the AquaCrop model, the model has been validated by using data from the experimental field for 2012. Part of the monitored data (full irrigation) were used for calibration of the model, while the remaining data (F1, F2, F3 and S1, S2, S3) were used to validate the model (Todorovic *et al.*, 2009). For simulation weather data, canopy development, sowing date, irrigation depths and soil characteristic were used as inputs.

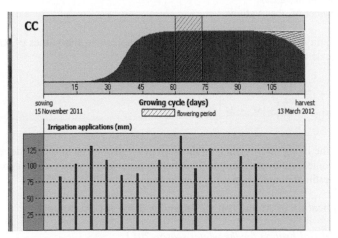

Figure 10.4. Calibration of the model for the full irrigaion treatment in the first year

10.5   Simulation with the AquaCrop model

Simulation models are designed to imitate the behaviour of the systems. Crop growth models contain a set of equations that estimate the production rate of biomass from the normal resources such as carbon dioxide, solar radiation and water. However, three main crop growth modules can be distinguished: i) carbon driven; ii) radiation driven; iii) water driven (Steduto *et al.*, 2009). The assumption is that there is a linear relation between biomass, growth rate and transpiration in the water driven crop growth models through a water productivity parameter. This approach results in a less complex structure and reduces the number of input parameters (Hsiao *et al.*, 2009).

The initial canopy cover was estimated by the model based on spacing between plant and cultivation method. After entering phenological data such as date of emergence, maximum cover and senescence, the canopy cover expansion was estimated automatically. The calibration parameters of canopy cover included the canopy growth coefficient (CGC), canopy decline (CDC) and early senescence. Sufficient amount of fertilizers was added to the field such as urea and phosphate carbonate as the Agricultural Research Corporation recommended.

Assessment of the model performance

The model performance was evaluated by using statistical parameters as root mean square error (RMSE) and index of agreement (*d*). The following equations were used in this study to examine the applicability and accuracy to use the AquaCrop model under Gezira conditions.

$$RMSE = \sqrt{\frac{\sum_{i=1}^{n}(S_{obs,i} - M_{model,i})^2}{n}} \tag{10.3}$$

Where:
S_i = simulated values
M_i = measured values
N   = number of observations

The unit of RMSE is the same for both variables and the values of RMSE close to zero indicate the best fit of the model.

$$d = 1 - \frac{\sum_{i=1}^{n}(S_i - M_i)^2}{\sum_{i=1}^{n}\left(\left|S_i - \overline{M}\right| + \left|M_i - \overline{M}\right|\right)^2} \tag{10.4}$$

Where d is the index of agreement, a measure of relative error in the model estimate. It is a dimensionless number from 0 to 1.0 where 0 describes a complete disagreement and 1.0 that the estimated and observed values are identical (Willmott, 1985).

10.6   Application of AquaCrop model under different treatments

10.6.1 Model Calibration for the winter season 2011

The AquaCrop model estimates the yield from the daily transpiration, considering the crop physiological characteristics. The effects of water stress on stomata conductance, canopy senescence and leaf expansion are expressed through factors, which vary from 0 to 1. The model accurately simulates Sunflower seed yield for high yield values. That water stress had less effect on yield in the first year showed a tendency to overestimate the lowest yields. There was a good agreement between the measured and simulated seed yield in the model calibration test. Table 10.3 shows the results of simulated and measured values of seed yield under different irrigation levels, which shows that there was a good agreement between simulated and observed values with minimum and maximum deviation of -3.70 and 4.55% respectively. The simulated seed yield decreased with water depth. The model simulated seed yield excellently when the irrigation interval changed from weekly to 15 and 20 days intervals with a deviation that ranged between -4.17% and -4.76% after the flowering stage. Moreover, the simulated versus measured values for water productivity showed a fairly satisfactory agreement, under all irrigation treatments (Table 10.3). The poorest simulation results indicate that the model over-estimated water productivity.

Consequently, the data show that the AquaCrop model can be used to take decisions to which crop priority should be given when the irrigation water is limited. Moreover, the simulated and measured values of seed yield for all treatments were well correlated with ($r^2 = 0.83$) and are plotted in Figure 10.5. It shows that the seed yield simulated by AquaCrop was well matched in all irrigation treatments.

The values of RMSE and *d-index* of the measured and simulated results are given in Table 10.4 for calibrating seed yield and water productivity. It shows that the simulated and measured yield data agreed well with the lower RMSE values ranging between 5 and 16% and a higher index of agreement varying between 0.87 and 0.98.

Statistical indicators of RMSE and d suggested that the model can be used to highly reliably to assess yield under full and deficit irrigation. The results reveal that the water productivity simulated by the AquaCrop model was over-estimated with low values of d ranging between 0.27 and 0.64.

Table 10.3. Measured and simulated results of calibrated seed yield (t/ha) and water productivity (kg/m^3) during the first winter season 2011

Treatment	Yield (t/ha)			WP (kg/m^3)		
	Measured	Simulated	Deviation (%)	Simulated (WPs)	Measured (WPm)	Deviation (%)
W	3.1	2.9	-6.45	0.39	0.26	51.7
F1	2.7	2.5	-7.41	0.36	0.25	43.5
F2	2.4	2.3	-4.17	0.31	0.23	36.0
F3	2.2	2.0	-4.76	0.28	0.21	34.3
S1	2.7	2.6	-3.70	0.37	0.26	41.3
S2	2.4	2.5	4.17	0.31	0.22	39.6
S3	2.2	2.3	4.55	0.34	0.21	60.9

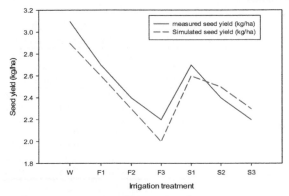

Figure 10.5. Irrigation treatments versus calibrated and measured yield (t/ha) for the first winter season 2011

Table 10.4. RMSE and index of agreement (d) for measured and simulated results of the calibrated seed yield and WP

Treatment	D	RMSE Yield (t/ha)	d	RMSE WP (kg/m^3)
W	0.94	0.15	0.46	0.14
F1	0.87	0.16	0.41	0.11
F2	0.94	0.05	0.27	0.11
F3	0.92	0.12	0.34	0.10
S1	0.89	0.08	0.64	0.13
S2	0.96	0.05	0.28	0.14
S3	0.98	0.05	0.36	0.12

In the AquaCrop model the water driven growth module uses the constant normalized water productivity for different climate conditions (both evaporative demand and the atmospheric CO_2 concentration), and it has a greater applicability at different locations under different settings.

10.6.2 Validation of the AquaCrop model

The values of RMSE and d- index of agreement for validated seed yield and water productivity for winter 2012 and 2013 seasons are presented in Table 10.5. The results revealed that the model simulated seed yield with a high degree of reliability, whereas the simulated water productivity was poor in the two seasons. The values of RMSE for the validation data ranged between 6 and 29% and the d-index between 0.78 and 0.95, which confirmed the model's simulation capability. The RMSE values of the validated water productivity ranged between 0.11 and 0.19 kg/m^3, and d values ranged from 0.28 to 0.64. This variation was caused because the d-index was calculated from three replicates, and there were significant differences in the simulated water productivities among them. General evaluation of the model depends on the RMSE values. The simulation is considered to be excellent when the RMSE is less than 10%, good when it is greater than 10% and less than 20%, fair if it is greater than 20% and less than 30% and poor if it is greater than 30% (Wallens *et al.*, (2005). The model validation was done by using the data for 2012 and 2013. Validation results are presented in Table 10.6. The calibration results show a good match between measured values and those simulated by the model.

Table 10.5. RMSE and d-index for measured and simulated results for validated seed yield (t/ha) and water productivity for the winter season 2012/2013

Treatment	D	RMSE	D	RMSE
		Yield		WP
		(t/ha)		(kg/m^3)
W	0.95	0.20	0.46	0.18
F1	0.89	0.29	0.59	0.11
F2	0.83	0.08	0.27	0.14
F3	0.95	0.18	0.34	0.13
S1	0.85	0.13	0.64	0.14
S2	0.78	0.06	0.28	0.19
S3	0.91	0.06	0.36	0.14

Actually, no significant differences were found between the calibrated and validated results. The differences between validated and measured seed yield under different irrigation levels deviated from -1.29 to 8.33% and -3.23 to 9.68% for the first and second season respectively. In contrast, the validated water productivity was slightly mismatched with the measured data because they were over-estimated in the simulated values. This might be due to the use of normalized water productivity in the model under field conditions.

The most significantly variation was noted in the simulated water productivity among all the irrigation treatments. The difference between simulated and measured values in case of water productivity ranged between 34.3 and 60.9% and between 6.06 and 56% for the first and second seasons respectively. The overestimated values were a result of the use of default values of coefficients water stress and normalized water productivity in the model.

Table 10.6. Measured and simulated results for the validated seed yield (t/ha) and water productivity for the winter seasons (2012 and 2013)

Treatment	Yield (t/ha)			WP (kg/m³)			Yield (t/ha)			WP (kg/m³)		
	Measured	Simulated	Deviation %	Measured	Simulated	Deviation %	Measured	Simulated	SD %	Measured	Simulated	Deviation %
	2012 season						2013 season					
W	3.1	3.1	-1.29	0.36	0.31	51.7	3.1	3.0	-3.23	0.28	0.39	39.3
F1	2.9	3.0	3.45	0.31	0.41	43.5	3.1	2.8	-9.68	0.35	0.42	20.0
F2	2.5	2.4	-4.00	0.29	0.41	36.0	2.7	2.6	-7.41	0.36	0.45	25.0
F3	2.1	2.2	4.76	0.26	0.37	34.3	2.3	2.4	4.35	0.25	0.39	56.0
S1	2.8	2.9	3.57	0.30	0.41	41.3	3.0	2.9	-3.33	0.34	0.40	17.7
S2	2.4	2.6	8.33	0.28	0.40	39.6	2.5	2.3	-8.00	0.33	0.35	6.1
S3	2.3	2.4	4.35	0.27	0.38	60.9	2.4	2.2	-8.33	0.32	0.37	15.6

Deviation%=(Simulated−measured)*100/measured.

10.6.3 Canopy cover development (%)

AquaCrop has also been used to predict canopy cover (CC) under different irrigation treatments. The calibrated canopy cover values under different irrigation treatments are shown in the Figures 10.6 and 10.7. The option in the model to estimate the initial canopy cover (CCo) from the sowing rate has been used. The canopy cover coefficient, canopy decline coefficient (CDC) and the stress indices for water stress affecting leaf expansion and early senescence are the most important canopy cover parameters. All these parameters and curve shapes have been manually changed and adjusted from the default values to reproduce the measured values of canopy cover. It was observed that there is a slight mismatch between the measured and simulated canopy cover in some crop stages. This is due to the slightly faster decline of the measured canopy cover compared with the simulated canopy cover. Therefore, it was observed that the calibrated canopy cover results under different irrigation levels were better and closer to the measured data.

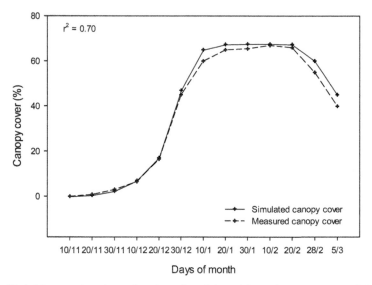

Figure 10.6. Measured and simulated results of the calibrated canopy cover (%) under full irrigation (W)

The performance values of RMSE and *d*-index for the simulated canopy cover for different irrigation treatments are presented in Table 10.7. As a result, the range of RMSE was between 1.9 and 10.1% and for the *d*-index from 0.96 to 0.99. Therefore, the AquaCrop model had a good agreement among the simulated and measured values of the canopy cover.

Figure 10.8 shows the simulated canopy cover for all treatments for two winter seasons. There were no significant differences between the two seasons. The large canopy cover was observed under the F1 treatments and there was a rapid decline in the stressed treatments, while there was a small canopy cover under weekly irrigation as simulated by the AquaCrop model. The simulated canopy cover was well simulated under optimal and mild water stress conditions.

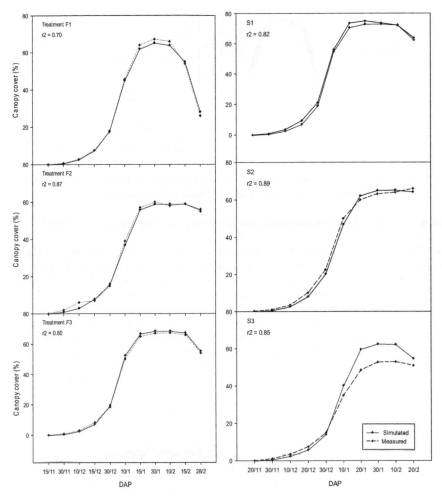

Figure 10.7. Measured and simulated results for the calibrated canopy cover (%) for all irrigation treatments during winter season 2011

Table 10.7. The RMSE and *d*-index for measured and simulated results for the calibrated canopy cover (%) for winter season 2011

Irrigation treatment	RMSE (%)	D
W	3.7	0.99
F1	5.5	0.99
F2	10.1	0.98
F3	3.2	0.97
S1	1.9	0.99
S2	2.2	0.99
S3	7.1	0.96

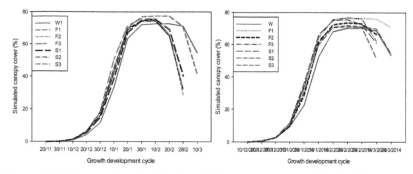

Figure 10.8. Simulated results for the validated canopy cover (%) for all irrigation treatments during the winter seasons 2012 and 2013

10.6.4   Model calibration and validation for the summer seasons 2012 and 2013

Sunflower parameterization has been done by using data sets of single locations (Todorovic *et al.*, 2009). Data from different climate and soil conditions are needed to investigate more for Sunflower. The performance of the calibrated data of the model was validated and evaluated by comparing the simulation results of yield and water productivity in the summers 2012 and 2013 with measured data. During these seasons the rainfall is an important attribute to this experiment. The model was calibrated by using field data from the 2012 growing season. Data from the full irrigation treatment as well as those under deficit irrigation treatments were used for calibration and validation with data for the summer season 2013. Also the air temperature data were used in the model to see the effect of temperature on the simulation results, but there was no significant effect of temperature on water stress (Figure 10.9).

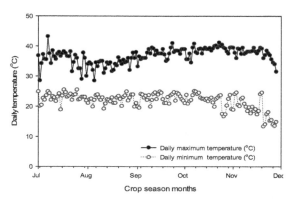

Figure10.9. Daily maximum and minimum temperature (°C) for summer season 2012

Based on the field measurements the Sunflower parameters time from sowing to emergence (6 days), to flowering (55 - 60 days) dependent on the irrigation treatments, and duration of flowering (9 - 11days) for different intra-row plant spacing and irrigation intervals were obtained. Water stress accelerated the flowering and seed filling duration, while under the full irrigation treatment the period was 3 - 4 days longer. The parameters used in the AquaCrop model to calibrate Sunflower under rainy conditions are presented in Table 10.8. The effect of soil fertility on yield was not addressed since sufficient amounts of urea were added in all treatments. On the other hand, the conservative parameters based on the default values were generally used for the three irrigation scenarios (Hsiao *et al.*, 2009). Moreover, biomass and harvest index data for the growing season were not available, such that a reliable calibration could not be performed by varying biomass and harvest index related parameters.

Table 10.8. Parameters used in the AquaCrop model for Sunflower calibration for summer season 2012

Parameter	Unit	Measured or calibrated
Base temperature	°C	10
Upper temperature	°C	35
Plant density (1)	Plants/m^2	4.2
Plant density (2)	Plants/m^2	3.1
Time to emergence	days	6
Time from sowing to maximum canopy	days	55 - 60
Canopy decline (CDC)	days	13 - 20
Time to senescence	days	75 - 85
Maximum canopy cover (CCx)	%	85
Duration of flowering	days	11
Maximum effective rooting depth	m	0.85
Day from sowing to maturity	days	100 - 110
Day from sowing to flowering	days	60
Water productivity (normalized)	g/m^2	17
Stomata stress coefficient curve shape	-	3
Upper threshold for canopy expansion	-	0.3
Lower threshold canopy expansion (P lower)		0.60
Canopy senescence stress coefficient (P upper)		0.55
Senescence stress coefficient curve shape	-	3.5
Leaf expansion stress coefficient curve shape		3.5
Saturated hydraulic conductivity	mm/day	15
Crop coefficient K_{cb}		1.10
Soil characteristics		Heavy clay

By adopting a trial and error approach, the Sunflower canopy cover growth was 18% as daily increase and a value of 10% decline per day for CDC during the late season. Some parameters were not recorded. In these cases the default values were used. Harvest index (HI) was not recorded during the study, but no increase in it was observed due to the negative impact of water stress on the harvest index occurred before flowering (Raes *et al.*, 2009). As there was no significant difference between the two varieties (Hysun 33 and Bohooth-1) in some characteristics such as phenological development and canopy cover, the difference was just in leaf colour; Bohooth-1 was having a light green colour. Thus the calibration and validation results seem to be similar by the model for Hysun 33 and Bohooth-1.

10.6.5 Results and discussion for the summer seasons 2012 and 2013

The AquaCrop model was calibrated and validated by using data for 2012 and 2013 summer seasons to predict yield, water productivity and canopy cover under different irrigation and agronomic treatments. Results show that the AquaCrop model simulated seed yield accurately in both seasons. The RMSE values for yield were small and the *d*-index values were very close to 1 under different irrigation treatments in both growing seasons. There was a slight mismatch in the measured yield values under deficit irrigation of 11.1 - 16.7 t/ha and 5.6 - 16.7 t/ha for calibration of the yield for Hysun33 and Bohooth-1 respectively. The lower RMSE calculated for all treatments in this study indicate a good agreement.

On the other hand, there was a trend for overestimate the water productivity under different treatments. The calibrated and validated water productivity was mismatched with measured data obtained from the experimental field (Table 10.9). Previous studies also reported that AquaCrop was overestimating water productivity for different crops. For instance, water productivity for Cotton was studied by Hussein *et al.* (2011) and for Sunflower by Ruzica *et al.* (2011).

Table 10.9. RMSE and *d*-index of agreement for measured and simulated results for the calibrated and validated seed yield (t/ha) and water productivity for Hysun 33 for two summer seasons 2012 and 2013

Treatment	Calibration				Validation			
	Yield (t/ha)		WP (kg/m^3)		Yield (t/ha)		WP (kg/m^3)	
	RMSE	*D*	RMSE	*D*	RMSE	*D*	RMSE	*d*
W1S1	0.04	0.99	0.06	0.84	0.03	0.99	0.03	0.79
W1S2	0.01	0.99	0.00	0.88	0.02	0.99	0.02	0.88
W2S1	0.04	0.89	0.09	0.36	0.05	0.97	0.02	0.78
W2S2	0.04	0.96	0.06	0.00	0.04	0.96	0.02	0.39
W3S1	0.12	0.93	0.15	0.00	0.04	0.98	0.01	0.30
W3S2	0.08	0.97	0.12	0.00	0.05	0.97	0.02	0.31

The calibrated model was validated by using data of 2013. The validation runs with calibrated data in AquaCrop showed good results for seed yield as indicated by RMSE and *d-index* values. The RMSE's and indices of agreement for Hysun33 and Bohooth-1 are shown in the Tables 10.9 and 10.10 for the calibration and validation data set. It was apparent that the simulated and measured values of seed yield were rather well in agreement for the two varieties. This means that differences in crop varieties had no significant effect on the calibration results. However, the performance indicator values of RMSE ranged between 0.01 and 0.12 t/ha and *d* varied between 0.89 and 0.99% for Hysun 33, and for Bohooth-1 they ranged between 0.06 and 0.17 t/ha with *d* from 0.64 to 0.99. The results show that the model performed well for simulating seed yield in the summer season. However, there was disagreement between the simulated and measured water productivity under severe water stress for the two crop varieties, because the *d* varied between 0.30 and 0.31 for Hysun33 and zero Bohooth-1 in the model validation.

Table 10.10. RMSE and *d*-index of agreement for measured and simulated results for the calibrated and validated seed yield (t/ha) and water productivity for Bohooth-1 for summer season 2012 and 2013

Treatment	Calibration				Validation			
	Yield (t/ha)		WP (kg/m^3)		Yield (t/ha)		WP (kg/m^3)	
	RMSE	*D*	RMSE	*D*	RMSE	*D*	RMSE	*d*
W1S1	0.12	0.98	0.05	0.77	0.12	0.94	0.01	0.96
W1S2	0.17	0.93	0.05	0.78	0.12	1.00	0.01	0.90
W2S1	0.06	0.93	0.02	0.82	0.04	1.00	0.00	0.44
W2S2	0.12	0.64	0.01	0.84	0.12	0.99	0.01	0.68
W3S1	0.17	0.92	0.00	0.00	0.04	1.00	0.04	0.00
W3S2	0.06	0.99	0.00	0.00	0.01	1.00	0.05	0.00

Araya *et al.* (2010a) simulated grain yield of Barley with deviation among observed data of -13 to 15.1% and for biomass from -4.3 to 14.6%. Araya *et al.* (2010b) also simulated Teff yield in Northern Ethiopia and they found good agreement among the simulated and observed grain yield and above ground biomass.

Tables 10.11 and 10.12 show the simulated and measured results for the summer season Hysun33 and Bohooth-1 by using the calibration data set for all treatments for seed yield and water productivity. The deviation of the simulated seed yield was -3.1 and 16.7% for Hysun33 and for Bohooth-1 -6.3 to -18.2% for the first season. For the second season the minimum deviation was observed for the Bohooth-1 seed yield (-1.1%), while the highest difference was 5.3%. The calibration and validation results for seed yield showed good agreement between measured and simulated values. The maximum difference was for the simulated water productivity for both varieties. Differences in calibrated and validated water productivity for Hysun 33 were 9.4 - 21.9% and -9.8 - 37.5% for 2012 and 2013 respectively. While the differences for calibrated and validated water productivity for Bohooth-1 were 6 - 34% and 3 - 35% for season 2012 and 2013 respectively.

On the other hand, several studies indicated AquaCrop model overestimated water use and water productivity for some crops under severe water stress in different climatic conditions (Evett and Tolk, 2009). Water productivity of Cotton was simulated by the model; the result indicated that there was over-estimation in simulated values compared to the measured ones (Farahani *et al.,* 2009). Moreover, another study was done to simulate water use and yield for Cotton by using the AquaCrop model in Spain. The results revealed that under severe water stress the model tended to over-estimate the water productivity due to the use of the constant normalized WP* in the model (García-Vila and Fereres, 2012). However, in this study AquaCrop over-estimated water productivity by 32.2 - 38% and by 30 - 35% for Hysun 33 and Bohooth-1 respectively under severe water stress in the second season.

The AquaCrop model was able to simulate canopy cover development accurately. The simulated and measured data for canopy cover under different treatments were normally distributed (Figure 10.10). The relationship between intra-row plant spacing and maximum canopy cover gave interesting results. The predicted maximum canopy cover data were observed in all wider intra-row plant spacing treatments. Irrigation at 10 days interval in both intra-row plant spacing gave maximum canopy cover and hence a higher actual seed yield. However, important seed yield reduction can be caused by reducing canopy growth due to water stress.

Figure 10.11 gives the simulated canopy covers for all treatments and represents the large canopy resulting in high seed yield and measured canopy cover for 2012. Severe water stress after crop establishment and at the end of the vegetative stages has induced fast senescence as well as decline of the canopy cover when compared with the non stressed treatments. The canopy cover for the stressed treatment W3 showed a rapid decline when compared with the full irrigated treatment (Figure 10.12).

Table 10.13 shows the results of model performance values for validated canopy cover under different irrigation intervals and two intra-row plant spacings. Where the results of statistical analysis of canopy cover trends for all the treatments are shown with an acceptable match between simulated and measured values. The minimum difference in simulated canopy cover was 0.2%, while the maximum difference was 5.4%.

Table 10.11. Measured and simulated results for the calibrated and validated seed yield and water productivity for Hysun 33 for two summer seasons (2012 and 2013)

Treatment	Yield (t/ha)			WP (kg/m³)			Yield (t/ha)			WP (kg/m³)		
	Calibration season 2012						Validation season 2013					
	Measured	Simulated	deviation (%)	measured	simulated	deviation (%)	measured	Simulated	deviation (%)	measured	simulated	deviation (%)
W1S1	3.2	3.1	-3	0.41	0.47	15	3.1	3.2	3	0.41	0.37	-10
W1S2	3.2	3.2	0	0.42	0.48	14	2.9	2.9	0	0.38	0.39	3
W2S1	2.7	2.6	-4	0.37	0.42	14	2.5	2.3	-8	0.44	0.54	23
W2S2	2.3	2.2	-4	0.34	0.38	12	2.2	2.1	-5	0.39	0.48	23
W3S1	1.8	2.1	17	0.32	0.39	21	1.8	1.7	-11	0.31	0.41	32
W3S2	1.8	2.0	11	0.32	0.41	28	1.9	1.7	-11	0.32	0.44	38

Table 10.12. Measured and simulated results for calibrated and validated seed yield and water productivity for Bohooth-1 for two summer seasons (2012 and 2013)

Treatment	Yield (t/ha)			WP (kg/m³)			Yield (t/ha)			WP (kg/m³)		
	Calibration season 2012						Validation season 2013					
	measured	Simulated	deviation (%)	measured	simulated	Deviation (%)	Measured	simulated	deviation (%)	measured	simulated	deviation (%)
W1S1	3.2	3.0	-6	0.41	0.49	20	3.1	2.8	-10	0.41	0.43	5
W1S2	3.2	2.9	-9	0.41	0.49	20	2.9	2.6	-10	0.39	0.41	5
W2S1	2.6	2.7	4	0.39	0.43	10	2.1	2.2	5	0.37	0.38	3
W2S2	2.2	2.4	9	0.33	0.36	6	2.2	1.9	-14	0.38	0.41	8
W3S1	1.8	2.1	18	0.32	0.43	34	1.9	2.0	5	0.33	0.43	30
W3S2	1.8	1.9	6	0.33	0.42	27	1.9	1.88	-1	0.33	0.44	35

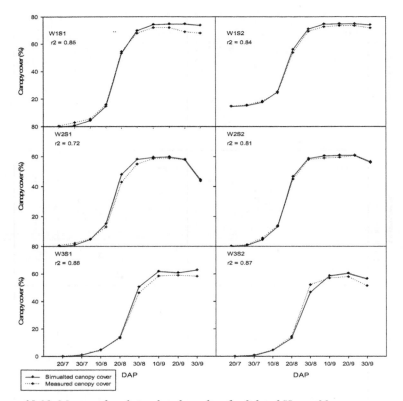

Figure 10.10. Measured and simulated results of validated Hysun 33 canopy cover (%) for all irrigation treatments in summer season 2013

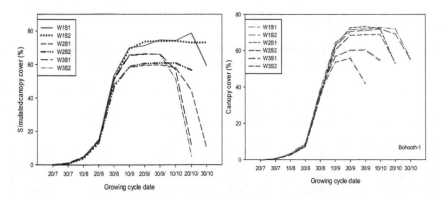

Figure 10.11. Simulated results of calibrated canopy cover (%) for two hybrids for all irrigation treatments in summer season 2012

The validation results show a reasonably close match between measured and simulated values by the model. The results also show that the model acceptably simulates the values of crop canopy under deficit irrigation. These results confirm that yield production can be predicted by simulating canopy cover percentage by the model.

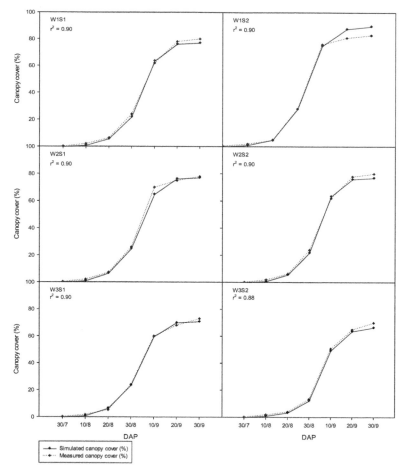

Figure 10.12. Measured and simulated results of validated Bohooth-1 canopy cover (%) for all irrigation treatments in summer season 2013

Table10.13. The RMSE, *d*-index and relative error percentage for the measured and simulated results for validated canopy cover (%) for Hysun 33 for summer season 2013

Treatment	Simulated	Measured	SD (%)	RMSE	D
W1S1	45.1	42.8	5.4	3.7	0.99
W1S2	41.1	39.3	4.5	6.2	0.99
W2S1	36.0	36.9	-2.6	2.1	0.99
W2S2	35.2	36.1	-2.5	1.6	0.99
W3S1	33.2	33.2	0.2	1.2	0.99
W3S2	28.2	29.3	-3.9	1.5	0.99

10.6.6 Model calibration and validation for the winter seasons 2012 and 2013

The AquaCrop model was used to simulate yield and water productivity for the winter season under different irrigation levels and to examine at which climatic conditions the model works well and accurately. In this way, the model was used the same way as has been done in the model simulation for summer crops. The same crop parameters were used to calibrate the model for two seasons of 2012 and 2013 except the climate data (ETo) for 2012/2013. The simulation results for calibrated and validated seed yield and the statistical test of the model are presented in the Tables 10.14 and 10.15. There were no significant differences between the simulated and measured values of seed yield. The differences between simulated and measured seed yield values were ranged between -6 - 11% and -21 - 15% in the first and second season respectively. The performance test shows the values of RMSE and d-index. The lower corresponding RMSE values were 0.07 - 0.14 and 0.11 - 0.29 and the higher d-index of 0.86 - 1.00 and 0.42 - 0.93 confirms that the model simulated the seed yield accurately. The simulated seed yield was well simulated under full and stress irrigation treatments. Despite there was slightly underestimation of the simulated seed yield under full and moderate water treatment, also there was a slight overestimation of the simulated seed yield under severe water stress treatments. Sunflower seed yield was improved on the model under water stress. To a certain degree, it was possible to adjust the yield by changing the water stress parameters to better fit the simulated values with the measured.

Table 10.14. The RMSE, d-index and relative error percentage for measured and simulated results for calibrated seed yield (t/ha) for Hysun 33 winter 2012

Treatment	Simulated	Measured	SD (%)	RMSE	D
W1S1	3.0	3.2	-6.25	0.14	0.93
W1S2	3.1	3.3	-6.06	0.14	0.96
W2S1	3.0	3.1	-3.23	0.07	0.98
W2S2	2.8	2.9	-3.45	0.07	0.86
W3S1	2.1	1.9	10.53	0.14	0.98
W3S2	2.0	1.9	5.26	0.07	1.00

Table 10.15. The RMSE, d-index and relative error percentage for the measured and simulated results for validated seed yield (t/ha) for Hysun 33 winter 2013

Treatment	Simulated	Measured	SD (%)	RMSE	D
W1S1	2.8	3.2	14.64	0.29	0.73
W1S2	2.9	3.3	13.45	0.28	0.42
W2S1	3.1	2.9	-7.74	0.17	0.93
W2S2	2.9	3.1	5.17	0.11	0.91
W3S1	2.2	1.8	-16.82	0.26	0.83
W3S2	2.0	1.6	-20.50	0.29	0.90

Tables 10.16 and 10.17 show the results of simulated and observed values for water productivity under different irrigation levels and the statistical performance for the first and second season. The differences were higher in the simulated water productivity and ranged from 7 to 39% and from 10 to 42% for calibrated and validated water productivity in the first and second season respectively. The simulated water productivity values under full irrigation were to some extent close to the measured values, but there was a larger mismatch between the simulated and measured values of water productivity in the two seasons under the severe water stress treatments. The reason could be attributable to the lower normalized water productivity value obtained by the model, which influenced by the water stress and thus lead to decrease in rate of transpiration. On the other hand, recommended Nitrogen was applied at flowering,

which increased transpiration and thus water productivity (plants absorb Nitrogen well under irrigation conditions). Raes *et al.* (2009) who indicated that lower application of Nitrogen may lead to lower water productivity. Under water stress the growing period was short because the plant synthesis accumulation is a function of time and many other environmental factors. However, water stress after the vegetative stage or before heading has induced fast senescence when compared to full irrigation treatments.

Table 10.16. The RMSE, *d*-index and relative error percentage for the measured and simulated results for calibrated water productivity (kg/m^3) for Hysun 33 winter 2012

Treatment	Simulated	Measured	SD (%)	RMSE	D
W1S1	0.41	0.35	17.8	0.06	0.00
W1S2	0.43	0.36	20.2	0.06	0.00
W2S1	0.52	0.49	6.5	0.10	0.99
W2S2	0.50	0.46	8.6	0.08	0.96
W3S1	0.39	0.30	28.6	0.09	0.00
W3S2	0.40	0.29	38.9	0.11	0.00

Table 10.17. The RMSE, *d*-index and relative error percentage for the measured and simulated results for validated water productivity (kg/m^3) for Hysun 33 winter 2013

Treatment	Simulated	Measured	SD (%)	RMSE	d
W1S1	0.39	0.34	14.5	0.05	0.44
W1S2	0.40	0.36	10.0	0.06	0.85
W2S1	0.47	0.33	42.2	0.12	0.00
W2S2	0.46	0.36	27.9	0.10	0.59
W3S1	0.41	0.32	27.3	0.11	0.00
W3S2	0.38	0.28	35.4	0.14	0.00

Figure 10.13 shows the simulated seed yield by the model plotted against the measured seed yield under different irrigation levels, which shows clearly the good agreement between model results and the data from the experiments. The simulated seed yield correlated strongly with the measured data with coefficient of determination r^2 of 0.99 and 0.86 for the first and second season respectively. Therefore, the model simulated seed yield with a high degree of reliability.

Figure 10.14 shows the simulated results for calibrated and validated water productivity under different irrigation levels and intra-row plant spacing. This was because the model did not simulate winter seed yield well, as well as it did summer seed yield, resulting in an overestimation of water productivity. The simulated water productivity did fit well with the measured in the first season, while in the second season there was a high variation under mild and stressed treatments.

10.7   Concluding remarks

AquaCrop was used with limited input data obtained from the field experiments and was relatively easy to use with a high degree of reliability.

The model performance was tested with very satisfactory results. AquaCrop was able to precisely simulate seed yield with RMSE (0.14) and index of agreement (*d*) less than 10%, values for the first season showed a good fit to the measured seed yield with simulated seed yield.

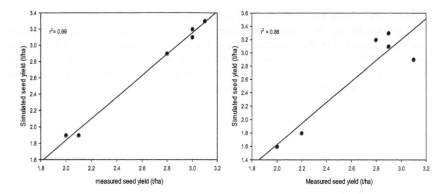

Figure 10.13. Measured and simulated seed yield (t/ha) for full and deficit irrigation for two seasons 2012 and 2013

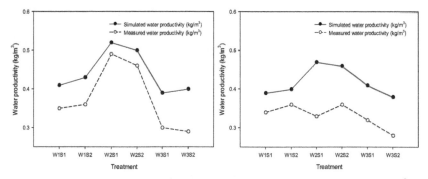

Figure

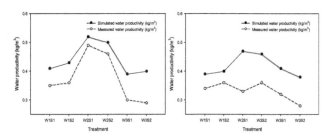

Figure 10.14. Comparison of measured and simulated water productivity (kg/m³) for full and deficit irrigation 2012/2013

The model was less satisfactory to predict water productivity under different irrigation intervals with mismatched measured water productivity, which ranged between 0.2 and 0.26 kg/m³ and simulated values varied between 0.28 and 0.39 kg/m³ when calibrated in the first season under water stress and full irrigation treatments respectively. However, AquaCrop had a tendency to over-estimate water productivity in all treatments.

Moreover, in some seasons the *d*-index varied between 0.44 and 0.00 for the simulated water productivity under different irrigation treatments, which indicated that there was disagreement between observed and simulated values.

The results of the simulated water productivity under different irrigation treatments revealed that the AquaCrop model was not accurate in this case. Under severe water deficit the simulated water productivity was significantly different. Thus, as also several previous studies indicated it needs more work and studies for more reliability of AquaCrop under certain conditions.

11 Evaluation and recommendations

11.1 Assessment of the existing irrigation system

The Gezira Scheme, located between the Blue Nile and the White Nile in Sudan, is one of the oldest and largest (880,000 ha), gravity irrigation scheme under one management in the world. It was established in 1925 to irrigate 126,000 ha and thereafter expanded with the Managil extension to about 880,000 ha. It receives water from the Sennar Dam, which supplies over 6 billion cubic metres from the Blue Nile and is divided into some 120,000 tenancies (Plusquellec, (1990b). Farmers operate the scheme in partnership with the government and the Sudan Gezira Board. The scheme has played an important role in the economic development of Sudan, serving as a major source of foreign exchange earnings and of Government revenue. However, due to changes in agriculture policies during the past ten years Gezira Scheme has suffered many problems with respect to irrigation management. Moreover, the expansion of the irrigated land in the scheme and the lack of maintenance of the irrigation canal system have caused inefficient use and water delivery. Inequity problems were found between the field outlet pipes and the headwork in the same canal system. Nevertheless, the quality of services provided to different parts of the Gezira irrigation system can be achieved by equity in water supply to each part during the season. In contrast, this scheme has been established more than 80 years ago and the irrigation system and practices still are applied without improvement. Due to ageing of the irrigation system, sedimentation, weed growth, water mismanagement and increased demand, water delivery at the farm level has become unreliable. Farmers tend to irrigate their fields (private crops) during the night.

The Gezira Scheme distributes water of the Blue Nile through a 4,300 km (2,700 miles) network of canals and ditches to irrigate the field crops. The irrigation water supply starts from the Sennar Dam to the main canal, then to major canals and from these canals to the minor canals. Thereafter the water flows by gravity through small canals, named AbuIshreens, which irrigate farmer's fields through a small ditch called AbuSitas.

During the past three decades irrigation in the scheme significantly changed from central management to self-management. The farmers have the responsibility to manage their water by themselves through water users associations. With expected change in agricultural policies and expansion of cultivated area in the scheme, water shortage will be one of the major challenges that will be faced. This needs a call for more efficient use of water supply. During the growing seasons, there were many problems and breaches caused by farmers. One of these breached canals has been closed at the mid season in the peak of the growing season, which led to water shortage in some part of our research area and losses in harvestable yields.

Some parts of the scheme experience characteristic drought during the summer crop season, whereas irrigation water is abundantly available during the winter. The Gezira Scheme achieves low water productivity in comparison to many irrigation schemes of the globe. Low productivity is mainly due to weak distribution and irrigation mismanagement

The water supply in Gezira Scheme is highly reliable at the beginning of the season. This leads farmers to fully irrigate their fields. Accordingly, productivity in the Gezira Scheme is very low with an average of 0.21 - 0.26 kg/m^3, due to reasons that include a delay of land preparation, inequity of water distribution and lack of awareness about irrigation scheduling among farmers, which is a major problem in Gezira Scheme. Due to this, there are delays in all other agricultural practices as sowing, fertilization and weeding, which results in low yields and less output. The current policies of subsidies in the scheme force the farmers to obtain a maximum yield from the irrigation water rather than maximum water productivity.

It is noteworthy that for the Gezira Scheme there is a potential to increase the value of irrigation water by efforts of saving water. Therefore, improved irrigation water management is the key to increase crop production and to improve water sharing and water conservation.

Since the Gezira Act of 2005 the responsibility of the water users associations is high. This includes water demand, freedom of water access, in addition to flexibility in canal cleaning and maintenance. However, at this time implementation of the act faces many constraints and therefore the act needs to be modified for better water management in the scheme. In addition, it will undoubtedly be an effort to change the Gezira Act 2005 in order to improve the water productivity. Accordingly, the quality of water supply to different parts of the scheme can be satisfied by equity in water distribution during the growing season. On the other hand, one of the major national goals is transforming the Gezira Scheme into a very significant contributor to increasing the hard currency reserve of Sudan and to accelerate the national economic growth and political stability. To realize this goal, introducing high value crops based on a detailed investigation of their crop and water productivity and profitability under local conditions is imperative. The crop productivity strongly depends upon the performance of the on-farm irrigation systems. Accordingly, crop productivity can be improved through proper irrigation scheduling and over-irrigation practices would have to be avoided. The overall, water availability is higher in the scheme than the demand, but due to mismanagement in water distribution and expansion of the cultivated area, the scheme would face water stress during the growing season, which would need more attention for enhancing water productivity.

11.2   Assessment of AquaCrop model in simulating yield under Gezira conditions

To describe the relationship between crop yield and water stress and to evaluate the water productivity, yield response to water has been used as a result of insufficient water supply of rainfall or irrigation during the growing seasons. Doorenbos (1979) used an empirical production function to assess the yield response to water stress. Simulation models that identify the effects of water on yield at the on-farm level may be a valuable approach in irrigation water management. There are many models for irrigation scheduling management and decision making such as CROPWAT, AquaCrop and DSSAT. In recent years there has been an increasing concern in crop production assessment and monitoring by using crop modelling programs. As such there is no specific model of water management that would ensure best results in terms of sustainability and enhancing water productivity. The current situation in the scheme is raising need to increase awareness about the crop modelling and to identify which model for crop production can best be applied.

It was found that it is laborious and expensive to study yield responses to different irrigation regimes in the field, or under more controlled conditions. Modelling techniques are therefore very useful for studying and developing new deficit irrigation strategies. Less consideration is given to crop models for water management purposes in Gezira Scheme. The AquaCrop model is a new model preferred for agricultural planning and water management, because of its simplicity and limited amount of data that is required. AquaCrop has been developed for a wide range of users such as water managers, Agricultural consultants and policymakers. Performance assessment of the model had so far not yet been addressed for crop productivity in Gezira Scheme.

In this research the performance of the AquaCrop model has been examined in order to simulate Sunflower growth under different irrigation levels at Gezira Scheme conditions. The model has been calibrated with data obtained from the experimental field trails under different irrigation scenarios associated with different agronomical practices with a high degree of agreement. In the absence of sufficient data from the experimental fields the data from other locations with the similar conditions have been used. The model calibration could be done by changing model parameters by using a trial and error approach and based on good matching between simulated outputs and measured data. In some cases, parameters such as upper and lower thresholds for canopy expansion, canopy senescence stress coefficient and upper threshold for stomatal closure, the recommended default values of the model have been used, due to difficult measurements in the field.

The simulated results corresponded very well with the measured values, which indicate that the AquaCrop model would be a useful tool for predicting yield under different irrigation levels at current climatic conditions.

In the AquaCrop model the default values of crop coefficient were higher than the measured values. These higher values could be related to the specific characteristics of the Sunflower variety of the research. Due to this, some parameters were overestimated, which led to inaccuracy in the model results, for example with respect to water productivity. However, under such conditions for better performance and accuracy more data would have to be collected. For more robust model calibration, it will be necessary to have data for more than two years of experimental fieldwork under different weather and soil conditions. Also variation in crop phenological properties among different cultivars can be further explored.

In this regard the AquaCrop model performance is considered reliable and acceptable for simulating seed yield and crop canopy under different water supply levels. However, the AquaCrop model would be a decision support tool to improve water management and increase water productivity and help decision makers and irrigation managers in future planning. In this context, the advantages of the AquaCrop model can be widely used with acceptable accuracy. It requires only available and simple field data and it allows easy simulation data that well match with measured data.

It will also be interesting to expand the model results to consider the results of plot experiments in research stations, or on-farm field trials with other crops. This would enable more analysis of the experimental data, which enables to combine research on crop productivity with water management in practice.

11.3   Recommendations for improvement

In this final section some recommendations will be presented. Some of them are derived from the findings of this research, while others need to be considered for future studies:

- Sunflower seed yield depends on many factors such as irrigation water management, soil fertility, sowing date and plant densities. For optimum seed yield an irrigation interval of 10 days and intra-row plant spacing of 40 cm are recommended under irrigated conditions;
- irrigation water status in Gezira Scheme need to be improved by reducing the irrigation frequency and amount of water supplied by skipping one or two irrigations depending on crop stage and crop type. Therefore, further studies would have to be focussed on the effect of deficit irrigation on different crops

and at the relevant crop stages;

- the adoption of deficit irrigation strategies is generally difficult and requires good knowledge. Irrigation interval of 20 days should be avoided for irrigating Sunflower during flowering and seed filling stages under Gezira conditions;

- the AquaCrop model would be a useful tool for water management strategies and planning with a limited field data for crop production under different irrigation strategies with various climatic conditions. However, model performance evaluation needs to be tested;

- within the scope of this study the effects of deficit irrigation strategies on the socio-economic sustainability of the agricultural communities and on water service quality in Gezira Scheme have not been addressed. Follow-up studies with respect to these aspects are recommended;

- studies dealing with the effect of various agronomic practices on Sunflower yield and yield components such as sowing date, row spacing and fertilizer application have not yet been considered. Such studies may result in a further improvement of the yield of Sunflower in Gezira Scheme.

Eventually water deficit is becoming an essential alternative irrigation strategy to increase water productivity and reduce agricultural water use in water scarce regions. To irrigate more closely to the crop water needs and additional irrigation during the critical growth stages is recommended. These recommended irrigation schedules can be easily handled by the local farmers and are valuable for increasing Sunflower production and its economic benefit.

12 References

Abdelhadi A. (2006) Calculation of crop water requirement for Sorghum under Gezira Condition Personal commuincation

Abdelhadi A., Hata T., Tanakamaru H., Tada A., Tariq M. (2000) Estimation of crop water requirements in arid region using Penman-Monteith equation with derived crop coefficients: a case study on Acala cotton in Sudan Gezira irrigated scheme. Agricultural Water Management 45:203-214.

Abdou S., El-Latif K.A., Farrag R., Yousef K. (2011) Response of sunflower yield and water relations to sowing dates and irrigation scheduling under middle Egypt condition. Adv. Appl. Sci. Res 2:141-150.

Adam H.S. (2014) Agroclimatology, Crop Water Requirement and Water Management. Water Management and Irrigation Institute,University of Gezira Book.

Ahmed M.E., Mahmoud F.A. (2010) Effect of irrigation intervals and inter – row spacing on yield, yield components and water use efficiency of sunflower. Journal of Applied Sciences Research 6:1446-1451.

Ali D., Naderidarbaghshahi M., Abdolmajid R., Majdnasiri R. (2013) Effect of Water Deficiency Stress on Yield and Yield Component of sunflower Cultivars in Asfahan. International Journal of Farming and Allied Sciemces 2:1319-1324.

Ali M.H., Hoque M.R., Hassan A.A., Khair A. (2007) Effects of deficit irrigation on yield, water productivity, and economic returns of wheat. Agricultural Water Management 92:151-161. DOI: http://dx.doi.org/10.1016/j.agwat.2007.05.010.

Ali M.H., Talukder M.S.U. (2008) Increasing water productivity in crop production—A synthesis. Agricultural Water Management 95:1201-1213. DOI: http://dx.doi.org/10.1016/j.agwat.2008.06.008.

Allen R.G., Pereira L.S., Raes D., Smith M. (1998) Guidelines for Computing Crop water Requirements. . FAO Irrigation and Drainage Paper no 56. Rome, Italy.

Andarzian B., Bannayan M., Steduto P., Mazraeh H., Barati M., Barati M., Rahnama A. (2011) Validation and testing of the AquaCrop model under full and deficit irrigated wheat production in Iran. Agricultural Water Management 100:1-8.

Araya A., Habtu S., Hadgu K.M., Kebede A., Dejene T. (2010a) Test of AquaCrop model in simulating biomass and yield of water deficient and irrigated barley (Hordeum vulgare). Agricultural Water Management 97:1838-1846. DOI: http://dx.doi.org/10.1016/j.agwat.2010.06.021.

Araya A., Keesstra S.D., Stroosnijder L. (2010b) Simulating yield response to water of Teff (Eragrostis tef) with FAO's AquaCrop model. Field Crops Research 116:196-204. DOI: http://dx.doi.org/10.1016/j.fcr.2009.12.010.

Ardakani M.R., Rahmati A., Yarnia M., Daneshiyan J.,Valizadeah M. (2006) Effect of plant density on seed yield and yield components of two sunflower hybrids. 9th Iranian congress of crop production and plant breeding. 9th Iranian congress of crop production and plant breeding. Aburayhan Pardis. Tehran University.P:9.

Asbag F.T., Gorttapeh A.H., Fayaz-Moghadam A., Saber-Rezaie M., Feghnabi F., Eizadkhah M., Jahannavard S., Parvizi S., Moghadam-Pour S.N. (2009) Effects of planting date and irrigation management on sunflower yield and yield components. Research Journal of Biological Sciences 4:306-308.

Ayana M. (2011) Deficit irrigation practices as alternative means of improving water use efficiencies in irrigated agriculture: Case study of maize crop at Arba Minch, Ethiopia. African Journal of Agricultural Research 6:226-235.

Bashir M.A.A.M., Younis M. (2014) Evaluation of full and deficit irrigation on two Sunflower hybrids under semi-arid environment of Gezira, Sudan. journal of Agri-Food and Applied Sciences 2:53-59.

Bos M.G., Kselik R>A>L., Allen R.G., Molden D. (2009) Water requirements for irrigation and the environment. Springer, New York.

Canavar Ö., Ellmer F., Chmielewski F. (2010) Investigation of yield and yield components of sunflower (Helianthus annuus L.) cultivars in the ecological conditions of Berlin (Germany). Helia 33:117-130.

Central Intelligence. Agency. (2015) The world factbook, Library/publication. Available at www.cia.gov.

Chimenti C.A., Pearson J., Hall A.J. (2002) Osmotic adjustment and yield maintenance under drought in sunflower. Field Crops Research 75:235-246. DOI: http://dx.doi.org/10.1016/S0378-4290(02)00029-1.

D'Andria R., Chiarandà F.Q., Magliulo V., Mori M. (1995) Yield and soil water uptake of sunflower sown in spring and summer. Agronomy Journal 87:1122-1128.

Doorenbos J., Kassam A.H. (1979) Yield response to water. Irrigation and Drainage Paper No.33, FAO,Italy:, pp 193.

Elias E., Salih A., Alaily F. (2001) Cracking patterns in the Vertisols of the Sudan Gezira at the end of dry season. International agrophysics 15:151-156.

Eltigani E.B.A., Adeeb A.M. (2014) Impact of changing policies on agricultural productivity: a case of the Gezira scheme. Sudan. J. Sustainable Agricultural Management and Informatics X.

English M., Raja S.N. (1996) Perspectives on Deficit Irrigation. agric Water Management 32:1-14.

Erdem T., Delibas L., Orta A.H. (2001) Water use characteristics of sunflower (Helianthus annuus L.) under deficit irrigation. Pakistan Journal of Biological Sciences 4:766-769.

Evett S.R., Tolk J.A. (2009) Introduction: Can water use efficiency be modeled well enough to impact crop management? Agronomy Journal 101:423-425.

Farahani H.J., Izzi G., Oweis T.Y. (2009) Parameterization and evaluation of the AquaCrop model for full and deficit irrigated cotton. Agronomy Journal 101:469-476.

Farahvash F., Mirshekari B., Seyahjani E.A. (2011) Effects of water deficit on some traits of three sunflower cultivars. Middle-East Journal of Scientific Research 9:584-587.

Farbrother H.G. (1996) Water Management options in the Gezira: A Review. Chatham Natural Resources Institute. U.K, p. pp 43.

Flagella Z., Rotunno T., Tarantino E., Di Caterina R., De Caro A. (2002) Changes in seed yield and oil fatty acid composition of high oleic sunflower (Helianthus annuus L.) hybrids in relation to the sowing date and the water regime. European Journal of Agronomy 17:221-230. DOI: http://dx.doi.org/10.1016/S1161-0301(02)00012-6.

Food and Agriculture Organisation of the United Nations (FAO). (2008) Statistical Databases. Available online at: http://faostat.fao.org/site/567/Desktop Default.aspx?PageID=567.

Food and Agriculture Organisation of the United Nations (FAO). (2011). Farming Systems Report, Synthesis of the Country Reports at the level of the Nile Basin. Rome, Italy.

Food and Agriculture Organization of United Nations (FAO) and Government of Sudan. (2011). Crop and Food Security Assessment Mission (CFSAM). Special report 2011. Rome, Italy.

García-Vila M., Fereres E. (2012) Combining the simulation crop model AquaCrop with an economic model for the optimization of irrigation management at farm level. European Journal of Agronomy 36:21-31. DOI: http://dx.doi.org/10.1016/j.eja.2011.08.003.

Geerts S., Raes D. (2009) Deficit irrigation as an on-farm strategy to maximize crop water productivity in dry areas. Agricultural Water Management 96:1275-1284.

Ghani A., Hussain M., Qureshi M.S. (2000) Effect of different irrigation regimens on the growth and yield of sunflower. International Journal of Agriculture and Biology 2:334-335.

Göksoy A., Demir A., Turan Z., Dağüstü N. (2004) Responses of sunflower (< i> Helianthus annuus</i> L.) to full and limited irrigation at different growth stages. Field Crops Research 87:167-178.

Guvele C.A., Featherstone A.M. (2001) Dynamics of irrigation water use in Sudan Gezira scheme. Water Policy 3:363-386. DOI: http://dx.doi.org/10.1016/S1366-7017(01)00077-0.

Hajhassani-Asl N.M.R., Ghafari M, Alizadeh A., Moradi-Aghdam A. (2009) The effecet of drought stress and defoliation on some agronomic traits, seed yield and yield componntes of sunflower (Alestar hibrid). Danesh- Novin Journal. 5:27-39.

Hamid R.M., Tavassoli A. (2013) Effect of Water Stress on Quantitative and Qualitative Characteristics of Yield in Sunflower (Helianthus annuus L.). Journal of Novel Applied Sciences 2:299-302.

Hsiao T.C., Heng L., Steduto P., Rojas-Lara B., Raes D., Fereres E. (2009) AquaCrop—The FAO crop model to simulate yield response to water: III. Parameterization and testing for maize. Agronomy Journal 101:448-459.

Huygen J., Van den Brook B.J., Kabat P. (1995) Hydra Model Triger, a soil water balance and crop growth submitted to ICID/FAO Workshop Sep. 1995,: Rome Irrigation Scheduling from Theory to Practice-FAO, Rome, Italy.

International Fund for Agricultural Development (IFAD). (1992) Eastern and southern Africa region, the Northern Province of Sudan Irrigation Rehabilitation Project. The Staff Appraisal Report, Khartoum, Sudan.

International Fund for Agricultural Development (IFAD). (2002). Executive Board – Seventy-Sixth Session,Country Strategic Opportunities Paper (COSOP). Rome, 5-6 September 2002.

Iqbal N., Ashraf M., Ashraf M. (2005) Influence of water stress and exogenous glycinebetaine on sunflower achene weight and oil percentage. International Journal of Environmental Science & Technology 2:155-160.

Kahlown M.A., Raoof A., Zubair M., Kemper W.D. (2007) Water use efficiency and economic feasibility of growing rice and wheat with sprinkler irrigation in the Indus Basin of Pakistan. Agricultural Water Management 87:292-298. DOI: http://dx.doi.org/10.1016/j.agwat.2006.07.011.

Kakar A.A., Soomro A. (2001) Effect of water stress on the growth, yield and oil content of sunflower. Pak. J. Agri. Sei. Vol 38:1-2.

Karam F., Lahoud R., Masaad R., Kabalan R., Breidi J., Chalita C., Rouphael Y. (2007) Evapotranspiration, seed yield and water use efficiency of drip irrigated sunflower under full and deficit irrigation conditions. Agricultural Water Management 90:213-223.

Katerji N., Campi P., Mastrorilli M. (2013) Productivity, evapotranspiration, and water use efficiency of corn and tomato crops simulated by AquaCrop under contrasting water stress conditions in the Mediterranean region. Agricultural Water Management 130:14-26.

Kazemeini S.A., Edalat M., Shekoofa A. (2009) Interaction effects of deficit irrigation and row spacing on sunflower (Helianthus annuus L.) growth, seed yield and oil yield. Afr. J. Agri. Res 4:1165-1170.

Kijne J.W., Baker R., Molden D. (2003a) Improving Water Productivity in Agriculture, Editor's overview: In: Kijne, J. W., Barker, R.M.D. (EDs), Water Productivity in Agriculture: Limits and Opportunities for Improvement. International Water Management Institute, Colombo, Sir Lanka:xi-xix.

Kijne J.W., Barker R., Molden D.J. (2003b) Water productivity in agriculture [electronic resource]: limits and opportunities for improvement CABI.

Lamm, F. R., Aiken, R. M., & Aboukheira, A. A. (2011). *Irrigation research with sunflowers in Kansas.* Paper presented at the Proc. 23rd Annual Central Plains Irrigation Conference. Burlington, CO, USA.

Mehmet D.K., Kolsarici O. (2011) Seed Yield and Oil Content of Some Sunflower (Helianthus annuus L.) hybrid Irrigation at Different Growth Stages. African Journal of Biotechnology 10:4591-4595.

Ministry of Agriculture and Forestry. (2008) Agriculture in Sudan. Annual report 2008.

Mobasser H.R., Tavassoli A. (2013) Effect of Water Stress on Quantitative and Qualitative Characteristics of Yield in Sunflower (Helianthus annuus L.).

Mohammad J.Z., Ghalavand A., Daneshian J. (2005) Effect of planting patterns of Sunflower on yield and extinction coefficient. Agron..Sustain. Dev. 25:513-518.

Mohammed S., Bakht J., Yousaf M., Amankhan M. (2013) Effect of Irrigation Regime on Growth and Seed Yield of Sunflower (Helianthus Annuus L.). Pak. J. Bot 45:1995-2000.

Molden D., Oweis T., Steduto P., Bindraban P., Hanjra M.A., Kijne J. (2010). Improving agricultural water productivity: between optimism and caution. Agricultural Water Management 97:528-535.

Morison J., Baker N., Mullineaux P., Davies W. (2008) Improving water use in crop production. Philosophical Transactions of the Royal Society B: Biological Sciences 363:639-658.

Mula Ahmed M.F., Ahmed Shouk A.K., Gasim Ahmed F. (2007) Effects of irrigation water quantities and seasonal variation on oil content and fatty acid composition of sunflower (Helianthus annuus L.). Journal of the Science of Food and Agriculture 87:1806-1809.

Omer A.M. (2008) Water resources and freshwater ecosystems in Sudan. Renewable and Sustainable Energy Reviews 12:2066-2091. DOI: http://dx.doi.org/10.1016/j.rser.2007.01.005.

Omer, M. A. E. (1991). Multicriterion approach to the evaluation on irrigation syatems performance *Thesis Submitted for the degree of Doctor of Philosophy, Uuiversity of Newcastle upon Tyne Department of Civil Engineering*

Oweis T., Hachum A. (2006) Water harvesting and supplemental irrigation for improved water productivity of dry farming systems in West Asia and North Africa. Agricultural Water Management 80:57-73.

Payero J., Melvin S., Irmak S. (2005) Response of soybean to deficit irrigation in the semi-arid environment of west-central Nebraska. Transactions of the ASAE 48:2189-2203.

Pereira L.S., Oweis T., Abdelaziz Z. (2002) Irrigation management under water scarcity. Agricultural Water Management 57:175-206.

Playán E.M., Luciano. (2006) Modernization and optimization of irrigation systems to increase water productivity. Agricultural Water Management 80:100-116.

Plusquellec H. (1990a) The Gezira Irrigation Scheme in Sudan Objectives, Design, and Performance. World Bank Technical Paper Number 120.

Raes D., Steduto P., Hsiao T.C., Fereres E. (2009) AquaCrop The FAO Crop Model to Simulate Yield Response to Water: II. Main Algorithms and Software Description. Agronomy Journal 101:438-447.

Rodrigues G.C., Pereira L.S. (2009) Assessing economic impacts of deficit irrigation as related to water productivity and water costs. Biosystems engineering 103:536-551.

Sedghi M., Sharifi R.S., Namvar A., Khandan-e-Bejandi T., Molaei P. (2008) Responses of sunflower yield and grain filling period to plant density and weed interference. Research Journal of Biological Sciences 3:1048-1053.

Shayesteh T., Ghobadi M., Ghobadi M-E., Mohammadi.G., Honarmand S.J., Allahmoradi P. (2013) Effects of Irrigation Regimes on Grain Yield and Its Components in Sunflower (Helianthus annuus L.) as Second Crop. Bulletin of Environment, Pharmacology and Life Sciences 2:68-72.

Steduto P., Hsiao T.C., Raes D., Fereres E. (2009) AquaCrop - The FAO crop model to simulate yield response to water: I. Concepts and underlying principles. Agronomy Journal 101:426-437.

Steduto P., Hsiao T.C., Fereres E. (2007) On the Conservative behavior of biomass water productivity. Irrigation Science: 25: 189-207.

Steer B.T., Coaldrake P.D., Pearson C.J., Canty C.P. (1986) Effects of nitrogen supply and population density on plant development and yield components of irrigated sunflower (Helianthus annuus L.). Field Crops Research 13:99-115. DOI: http://dx.doi.org/10.1016/0378-4290(86)90014-6.

Stricevic R., Cosic M., Djurovic N., Pejic B., Maksimovic L. (2011) Assessment of the FAO AquaCrop model in the simulation of rainfed and supplementally irrigated maize, sugar beet and sunflower. Agricultural Water Management 98:1615-1621.

Sun H.-Y., Liu C.-M., Zhang X.-Y., Shen Y.-J., Zhang Y.-Q. (2006) Effects of irrigation on water balance, yield and WUE of winter wheat in the North China Plain. Agricultural Water Management 85:211-218.

Taha M., Mishra B., Acharya N. (2001) Effect of irrigation and nitrogen on yield and yield attributing characters of sunflower. Ann. Agric. Res 22:182-186.

Todorovic M., Albrizio R., Zivotic L., Saab M.T.A., Stöckle C., Steduto P. (2009) Assessment of AquaCrop, CropSyst, and WOFOST models in the simulation of sunflower growth under different water regimes. Agronomy Journal 101:509-521.

Unger P.W. (1983) Irrigation effect on sunflower growth, development, and water use. Field Crops Research 7:181-194.

Vazifedoust M., Van Dam J., Feddes R., Feizi M. (2008) Increasing water productivity of irrigated crops under limited water supply at field scale. Agricultural Water Management 95:89-102.

Vega C.R., Andrade F.H., Sadras V.O., Uhart S.A., Valentinuz O.R. (2001) Seed number as a function of growth. A comparative study in soybean, sunflower, and maize. Crop Science 41:748-754.

Willmott C.J., Akleson, G.S., Davis, R.E., Feddema, J.J., Klink, K.M., Legates, D.R., Odonnell, J. and Rowe, C.M. (1985) Statistic for the evaluation and comparison of models. J. Geophys. Res.:8995 9005. www.wikipedia.org/wiki/Sudan.

Yagoub S., Osman A., Abdesalam A. (2010) Effect of Watering Intrvals and Weeding on Growth and Yield of Sunflower (Helianthus annuus l). Journal of Science and Technology 11.

Zaroug M.G. (2006) Country Pasture/Forage Resource Profiles, Sudan. Food and Agriculture Organization for United Nations (FAO).

Zhang H. (2003) Improving Water productivity through Deficit Irrigation:example fromSyria, the North China Plain and Oregon, USA, In: Kijen, J.W., Barker, R.M.D.(EDs), Water produvtivity in Agriculture: Limits and opportunities for Improvement. International Water Management Institute, Colombo, Sir Lanka:301-309.

Zwart S.J., Bastiaanssen W.G.M. (2004) Review of measured crop water productivity values for irrigated wheat, rice, cotton and maize. Agricultural Water Management 69:115-133. DOI: http://dx.doi.org/10.1016/j.agwat.2004.04.007.

Annexes

ANNEX A. Symbols

B	Above ground biomass (t/ha)
BD	Bulk density
CC	Canopy cover (%)
CC_x	Maximum canopy cover (%)
CCo	Initial canopy cover
CDC	Crop canopy decline (day)
CGC	Canopy growth coefficient
CWR	Crop water requirements
d	D-index of agreement
D	Drainage from the crop root zone (mm)
Dp	Deep percolation
ea	Actual vapour pressure (kPa)
E	Evaporation from the upper soil layer (mm)
Es	Evaporation from soil or open area (mm)
ETa	Actual evapotranspiration (mm)
ETc	Evapotranspiration (mm)
ET_O	Reference evapotranspiration (mm)
ETp	Potential evapotranspiration (mm)
ew	Water use ratio
EWP	Economic water productivity $€\ m^{-3}$
FC	Field capacity
Fnet	Change in total water in the root zone from underground water movement
G	Soil heat flux density (MJ m^2/day)
Gs	Stomatal conductance
ha	hectare
HI	Harvest index
I	Irrigation (mm)
IWP	Irrigation water productivity, kg/m^3
IWU_{Farm}	farm irrigation water use (m^3)
Kc	Crop coefficient
K_{cb}	Basal crop coefficient, dimensionless
K_e	Soil evaporation coefficient, dimensionless
kg	kilogram
km	kilometer
Ks	Stress coefficient in AquaCrop model
K_s	Water stress coefficient, dimensionless
K_{sat}	Saturated hydraulic conductivity (mm day-1)
K_y	Yield response factor, dimensionless
P	Precipitation (mm)
PWP	Permanent wilting point
R	Runoff (mm)
r^2	Coefficient of determination
RMSE	Root mean square error
Rn	net radiation at the crop surface (MJ m^2/day)

SMC	Soil moisture content
T	Temperature
T	tonne
T	Crop transpiration (mm)
TAW	Total available water in the root zone (mm/m)
Tr	Crop Transpiration
TWU	Total water use (m^3)
WP	Water productivity (kg/m^3)
WP*	Normalized water productivity (g/m^2)
Y	Grain yield (t/ha)
Ya	Actual crop yield (kg/ha)
Yx	maximum and actual yield (kg/ha)
u2	wind speed at 2 m height (m/s),
Zr	rooting depth (m)
ΔSW	variation in soil water content between planting and harvesting (mm)
θ_{FC}	soil water content at field capacity (m^3/m^3)
θ_{WP}	soil water content at the wilting point (m^3/m^3)
γ	psychrometric constant (kPa/°C).

ANNEX B. Abbreviations

DAP	Day after planting
DI	Deficit irrigation
FAO	Food and Agricultural Organization for the United Nation
GDP	Gross domestic product
GRSF	Gezira Research Station Farm
MSL	Mean sea level
SD	Standard deviation
SDG	Sudanese Pound
SE	Standard error of means
UNDP	United Nation Development Programme
WUA	Water users association

ANNEX C. Units

1 feddan	$= 4200 \text{ m}^2 = 0.42 \text{ ha}$
1 ha	$= 10,000 \text{ m}^2$
ha	$= 2.38 \text{ fedan}$
mm	$= 10 \text{ m}^3/\text{ha}$
1 m³/ha/day	$= 0.1 \text{ mm/day}$

ANNEX D. Some information of soil chemical and physical analysis, irrigation scheduling for the Sunflower experiments and about the crop water requirement of main crops in Gezira and crop rotation of Gezira crops

Table D.1. Soil chemical and physical characteristics of the experimental field (2012)

Chemical Properties

Depth (cm)	EC (ds/m)	C/N	CaCO$_3$ (%)	N (%)	O.C (%)	Soluble cations (me/l)			Soluble Anions (me/l)		SAR	PH
						Na	Ca	Mg	Cl	HCO$_3$		
0 - 30	1.0	4	5.2	0.050	0.218	1.8	3.0	1.0	2.2	1.8	1	7.9
30 - 60	0.5	4	4.8	0.045	0.187	1.8	2.5	1.0	2.0	1.6	1	8.0
60 - 90	1.8	5	5.0	0.040	0.187	2.57	4.0	1.5	2.0	1.7	7	7.9

Physical Properties

Depth (cm)	Clay (%)	Silt (%)	Sand fine (%)	Sand coarse (%)	BD (g/cm^3)	FC (%)	WP (%)	OM (%)	H.C (Cm/hr)
0 - 30	59	28	5	8	1.60	38.2	20.7	0.34	4.2
30 - 60	67	22	6	5	1.52	45.9	24.9	0.31	2.61
60 - 90	59	28	5	8	1.78	41.9	22.8	0.13	0.22

Table D.2: Example of irrigation scheduling for Sunflower under Gezira clay conditions

Irrigation treatment	1st	2nd	3rd	4th	5th	6th	7th	m^3/fed	Average irrigation (m^3/fed)	m^3/ha	Average irrigation (m^3/ha)
W1S2	699	464	416	471	473	473	479	2780	463	6610	1100
Mm	166	111	99	112	113	112	114	661			
W1S1	884	466	414	475	480	478	463	2780	462	6600	1100
Mm	210	111	98	113	114	114	110	660			
W2S1	816	469	424	478	472	474		2320	463	5510	1100
Mm	194	112	101	114	112	113		551			
W2S2	896	479	421	472	485	453		2310	462	5500	1100
Mm	213	114	100	112	115	108		550			
W3S1	940	479	462	486	482			1910	477	4540	1140
Mm	224	114	110	116	115			454			
W3S2	907	477	447	480	480			1880	471	4480	1120

Table D.3. Calendar of cultivation practices during the three growing years for experiment One

Event	2011	2012	2013
Sowing date and first irrigation	15/11/2011	19/11/2012	5/12/2013
Thinning (DAS)	19	17	20
Start of irrigation treatment (DAS)	60	58	55
Last Irrigation (DAS)	90	85	83
Physiological maturity (DAS)	110	112	97
Harvest (DAS)	125	127	112

Table D.4. Calendar of cultivation practices for experiment Two throughout the growing season

Practices	2012	2013
Sowing date and first irrigation	19/11/2012	5/12/2013
Weeding and/or thinning (DAS)	14	20
Start of irrigation treatment (DAS)	27	24
Last Irrigation (DAS)	82	85
Physiological maturity (DAS)	95	90
Harvest (DAS)	115	110

Table D.5. Crop water requirement for the main crops in Gezira Scheme
Peak daily water requirement (m^3/ha/day)

Month	Cotton LS	Cotton MS	Ground -nut	Wheat	Sorghum	Fruit & vegetable	Fodder	Monthly total
Jan	55.7					57.1		113
Feb	50.0			19.8		67.4		137
Mar	8.1					76.4		85
Apr						85.2		85
May						90.4		90
Jun			95.7		63.5	94.0	87.8	341
Jul			28.6		20.7	53.6	29.8	133
Aug	14.3	10.7	37.6		40.9	37.4	33.8	175
Sep	28.8	54.7	58.5		59.5	54.0	40.0	296
Oct	67.8	90.7	30.9	19.8	33.3	65.9	34.7	343
Nov	70.4	86.9		52.1		60.0		269
Dec	63.1	50.9		56.6		54.7		225
Annual total	358	294	251	148	218	796	226	2290

(Source: FAO Nile country report, 2008)

ANNEX E. Water flow measurements

Water flow measurements were carried out by using the current meter method, which is based on determining the mean discharge by using the water velocity and cross-sectional area. The following equation is used to calculate the discharge:

$$Q = V.A. \tag{E.1}$$

Where:
Q   = discharge (m^3/s)
V   =velocity (m/s)
A   = cross-sectional area (m)

The procedure of the current meter is the use of the rotor or propeller to determine the velocity of the water at the point where the current meter is set. Where water velocity was calculated by using following equation;

$$V = 0.008 + 0.2667n \tag{E.2}$$

Where
n: number of pulses counts/ time in second $\qquad$ (E.3)

ANNEX F. Samenvatting

Als gevolg van de groeiende wereldbevolking, zal de landbouwproductie moeten toenemen. Toch neemt het deel van het zoete water dat beschikbaar is voor de landbouw (72%) af. Onder de huidige waterschaarste omstandigheden, zou het beperkte beschikbare water efficiënter moeten worden gebruikt. Eerder onderzoek in de landbouw was vooral gericht op het maximaliseren van de productie. De laatste jaren is het onderzoek meer gericht op het productie systeem, met name de beschikbaarheid van grond en water, waarbij het water de beperkende factor in de teelt is. In regio's waar waterschaarste in principe de beperkende factor voor de teelt is, zijn de boeren geïnteresseerd in het telen van gewassen die zich kunnen aanpassen aan droogte.

Het Gezira systeem is Soedan's oudste en grootste zwaartekracht irrigatiesysteem in Soedan. De totale oppervlakte van het Gezira systeem in de centrale klei vlakten is ongeveer 880.000 ha, gelegen tussen de Blauwe Nijl en de Witte Nijl. Het systeem heeft een belangrijke rol in de economische ontwikkeling van het land gespeeld, en is een belangrijke bron van deviezen. Het heeft ook bijgedragen aan de nationale voedselzekerheid en in het genereren van een inkomen voor de 2,7 miljoen mensen die nu in het systeem leven. In het systeem zijn de boeren niet de eigenaar van de grond, ze zijn pachters. Het gebied is verdeeld over 120.000 pacht overeenkomsten met een gemiddelde oppervlakte van ongeveer 7,3 ha. De originele maat voor een gewone pacht in het oude Gezira systeem was 16,8 ha, maar een aantal van de pachten zijn door de jaren heen omgebouwd tot half-pacht overeenkomsten. De deelname van pachters in de landbouw is over meer dan twee generaties gedaald door de stijgende levensstandaard en beter onderwijs. Er wordt geschat dat meer dan de helft van alle pachters nu niet meer betrokken zijn in de landbouw en de meesten van hen die nog steeds actief zijn, hebben ook een baan erbuiten.

Het Gezira systeem werd beheerd door de Sudan Gezira Board (SGB) als een staatsbedrijf. De SGB werkte nauw samen met het Ministerie van Irrigatie en Water Voorraden (MOIWR), dat verantwoordelijk was voor de exploitatie, het onderhoud en het beheer van de hoofd, zij en secondaire kanalen (irrigatie systeem). De SGB was verantwoordelijk voor het landbouwkundige beheer. Het had irrigatie commissies langs de secondaire kanalen en vertegenwoordigers van elk van deze commissies vormden het irrigatie comité op blokniveau. In 2005 is de nieuwe wet voor de regeling van het beheer in werking gesteld door de SGB 'de Gezira Act 2005". De verantwoordelijkheid voor het beheer en het onderhoud van de zij- en secondaire kanalen werd toegewezen aan de SGB. Op basis van deze wet heeft de SGB 17.000 water gebruikers groepen opgericht, die verenigingen van water gebruikers (VWG) worden genoemd. De nieuwe verantwoordelijkheden van de VWGs betreffen het beheer en onderhoud van de lagere orde kanalen en betrokkenheid bij het beheer en onderhoud van de secondaire kanalen.

In het Gezira systeem maken de meeste boeren in sommige gebieden gebruik van de korte regentijd van juni tot september voor de teelt van gewassen zoals sorghum en groenten. De landbouw in het Gezira systeem wordt gedomineerd door de gewasproductie. De belangrijkste gewassen zijn sorghum, tarwe, aardnoten en het oliehoudende gewas sesam. Het schema is als volgt: van juni tot juli is het de tijd voor sorghum, katoen, aardnoten, sesam en groenten met behulp van vruchtwisseling en van november tot december voor de teelt van tarwe en groenten.

Zonnebloem (*Helianthus annuus L.*) is een belangrijk olie gewas in de wereld en in Soedan. Het is een nieuw eetbaar olie gewas in Soedan, de zaden hebben een oliegehalte van 40 - 50% en 30% verteerbaar eiwit en kunnen derhalve worden gebruikt als een bron van voedsel voor de mens of als pluimveevoer. Zonnebloem koek kan worden gebruikt als veevoer. Het kan worden gekweekt als winter gewas onder geïrrigeerde omstandigheden en als zomer gewas onder regenafhankelijke omstandigheden. Zonnebloem is uitgegroeid tot een belangrijk olie gewas voor zowel de boeren, als de consumenten in Soedan. Het is een gewas dat goed past in het lokale teeltsysteem en wordt beschouwd als de belangrijkste olie oogst van het land. Hoewel experimenten met Zonnebloemen in Soedan al in de jaren 1980 zijn begonnen, begon de commerciële productie begon laat en vooral gericht op regenafhankelijke teelt. De gecultiveerde gebieden bereikten in 1990 - 1991 123.000 ha met een gemiddelde opbrengst van 714 kg/ha. Recent zijn hoogproductieve hybriden geïntroduceerd en onder irrigatie geteeld. In Soedan wordt zonnebloem nu in twee seizoenen (zomer en winter) geteeld en wordt het erkend als een gewas met een hoog potentieel dat met succes kan voldoen aan de toekomstige olie-eisen. In het Gezira systeem was het areaal met Zonnebloemen voor de zomer en de winter van 2013 respectievelijk 2570 en 619 ha. Terwijl de gemiddelde opbrengst varieerde tussen 1,2 en 2,2 t/ha (0,5 - 0,9 t/feddan).

Zonnebloem is gecategoriseerd als een laag tot gemiddeld droogte gevoelig gewas. De droogte resistente eigenschappen kunnen worden toegeschreven aan een uitgebreid wortelstelsel, waarbij water en voedingsstoffen tot een diepte van 3 m kunnen worden onttrokken. Uit de beschikbare informatie blijkt dat de zaadopbrengst reactie op water meestal het grootst is wanneer zonnebloem wordt geïrrigeerd in een laat bloem ontluikend stadium vóór de bloei fasen.

Kennis betreffende de effecten van toediening van irrigatie water op de productie van zonnebloemen en de water productiviteit voor gewassen onder water schaarse omstandigheden wordt steeds belangrijker. Toediening van irrigatie water is bijzonder belangrijk omdat veel veldgewassen gevoeliger zijn voor watertekort tijdens specifieke fonologische stadia. In de plantaardige productie kan, in plaats van het bereiken van een maximaal rendement op basis van volledige irrigatie, de water productiviteit voor gewassen geoptimaliseerd worden binnen het concept van beperkte irrigatie. Beperkte irrigatie is een optimale strategie waarbij irrigatie wordt toegepast tijdens droogte gevoelige groeistadia van een gewas. Beperkte irrigatie is op grote schaal onderzocht als een waardevolle en duurzame productie strategie in droge gebieden. Door de beperking van water toediening tijdens droogte gevoelige groeistadia, is deze praktijk gericht op het maximaliseren van de water productiviteit voor gewassen en om daarmee de opbrengst te stabiliseren in plaats van te maximaliseren. Naast een correcte toepassing van beperkte irrigatie vereist dit een grondige beoordeling van het economische effect van de opbrengst vermindering veroorzaakt door droogte stress.

De zonnebloem heeft verschillende groeistadia: opkomst, vegetatieve groei, reproductie, bloei, zaad vorming en rijping. Watertekort in elk stadium leidt tot vermindering van de zaadopbrengst en het oliegehalte. De behandelingen die zijn uitgevoerd om het effect van watertekort in verschillende groeistadia te bestuderen toonden aan dat de zonnebloem wezenlijk wordt aangetast door watertekort dat zich in de gevoelige bloei, zaadvorming en rijping fasen voordoet. De hoogste opbrengst werd verkregen bij de behandelingen zonder watertekort tijdens deze fasen. Ook werd geconstateerd dat beperken van irrigatie water aan het begin van de bloeifase de zaadopbrengst met 25% verminderd door het verminderen van zowel het aantal zaden als het individuele gewicht van het zaad, terwijl in het midden van de bloeifase de zaadopbrengst minder werd beïnvloed door het beperkte irrigatie water. Er werd ook geconstateerd dat voor hoge zaadopbrengst zonnebloemen drie keer moeten worden geïrrigeerd met volledige of beperkte toediening van irrigatie water aan het begin, de bloeien en de zaadvorming fasen. Bij een beperkte beschikbaarheid van water, moet verminderde toevoer tijdens de bloeiperiode worden vermeden.

Zonnebloem is een korte duur gewas en kan goed in het huidige teeltsysteem worden ingepast zonder belangrijke veranderingen in de landbouwkundige exploitatie. Het patroon van planten speelt een belangrijke rol in het verhogen van de opbrengst van zonnebloemen. De bloei is de meest kritieke fase voor watertekort veroorzaakt door beperkte irrigatie, om deze reden moet beperkte irrigatie in de vroege en midden bloei stadia worden vermeden, terwijl beperkte irrigatie tijdens de zaadvorming acceptabel kan zijn wanneer er tekort is aan irrigatie water. Eerdere studies toonden aan dat de waterbehoefte van zonnebloemen afhankelijk van het klimaat en de lengte van de totale groeiperiode tussen 600 en 1000 mm varieert. Hoge verdampingswaarden treden op tijdens de zaadvorming en vroege rijpingsperiode. De waterbehoefte van zonnebloemen is relatief hoog in vergelijking met andere gewassen. Niettemin hebben ze de mogelijkheid om korte perioden een bodemvochtigheid tekort tot 15 atmosfeer te doorstaan.

Water gebruik door een gewas bij voldoende beschikbaar bodemvocht wordt voornamelijk beïnvloed door de kruin en de weersomstandigheden. Deze effecten worden vertegenwoordigd door de seizoensgebonden gewas coëfficiënt en de referentie gewasverdamping (ETo) van de atmosfeer. De gewas coëfficiënt geeft de fractie van de potentiële gewasverdamping (ETp) die het gewas naar verwachting gebruikt op een bepaalde dag. De waarde voor de gewas coëfficiënt verandert meestal met het groeistadium. Het effect van watertekort op de opbrengst en de opbrengst reactie factor voor zonnebloemen is zeer essentieel om de water productiviteit voor gewassen te beoordelen.

In het Gezira systeem is onderzoek naar het effect van watertekort op de opbrengst van zonnebloemen en water productiviteit voor gewassen schaars. In deze studie was de belangrijkste doelstelling het bepalen van de waterbehoefte van het gewas en de gewas coëfficiënten. De studie is uitgevoerd op de proefboerderij bij het Gezira systeem voor het bepalen van de waterbehoefte en de water productiviteit voor gewassen te bepalen en om het effect van beperkte irrigatie op de opbrengst van zonnebloemen en de opbrengst componenten onder Gezira kleigrond omstandigheden te onderzoeken. Er zijn drie experimenten uitgevoerd om de zaadopbrengst van zonnebloemen bij verschillende irrigatie intervallen te bepalen en twee testen met verschillende inter-rij plantafstand voor de winter en de zomer. Het eerste experiment is uitgevoerd om de gewas coëfficiënt (Kc) van zonnebloem en het effect van verschillende irrigatie intervallen op de opbrengst van zonnebloemen en de opbrengstcomponenten in het winterseizoen te bepalen, alsmede om het effect van watertekort op specifieke groeistadia te onderzoeken. Drie irrigatie intervallen zijn geselecteerd om de gevoeligheid van twee groeifasen (bloeifase en zaad vullen fasen) voor watertekort te onderzoeken. Irrigatie intervallen van 10, 15 en 20 dagen zijn getest op de fase van 50% bloei en de fase van zaad vullen, die zijn vergeleken met volledige irrigatie (wekelijks interval). Zonnebloem Hysun 33 (Hybrid) is geselecteerd om de water productiviteit voor gewassen en economische water productiviteit onder zes verschillende irrigatie behandelingen te evalueren. De evaluatie is uitgevoerd op basis van de zaadopbrengst verkregen uit de drie groeiseizoenen: het eerste seizoen begon op de 14 november 2011, het tweede seizoen op 19 november 2012 en het derde seizoen op het 5 december, 2013. Er kon worden geconcludeerd dat de hoogste opbrengst werd verkregen onder wekelijks irrigatie en onder 10 dagen intervallen na de bloei en zaad vullen fasen. Bij 20 dagen intervallen nam het rendement met meer dan 40% af.

Op basis van de wekelijkse irrigatie behandeling is de gewas coëfficiënt van zonnebloemen bepaald met behulp van de procedure die is aanbevolen door de Voedsel- en Landbouworganisatie van de Verenigde Naties (FAO). Deze procedure is gebaseerd op een berekening van het vochtgehalte voor het schatten van de gewasverdamping. Uit de meteorologische gegevens, verkregen van het Meteorologische Station dat het dichtst bij ons proefveld lag, is de referentie gewasverdamping (ETo) berekend met behulp van het FAO-programma EToCalc 2009 software. Uit de resultaten bleek dat de gewas coëfficiënt voor de begin fase, de ontwikkeling, halverwege het seizoen en in de rijping fasen respectievelijk 0,53, 1,1, 1,3 en 0,63 waren.

Bovendien is bij de wekelijkse irrigatie voor het eerste, tweede en derde jaar de maximale zaadopbrengst van respectievelijk 3130, 3140 en 3100 kg/ha verkregen. Irrigatie iedere 10, 15 en 20 dagen na de bloeifase verlaagde in het eerste seizoen de zaadopbrengst met respectievelijk 15, 23 en 34%, terwijl het in het derde seizoen 8, 20 en 31% was. Bovendien was de afname in de irrigatie behandelingen van 10, 15 en 20 dagen na de zaadvorming 10, 25 en 30% en 9, 24 en 26% in respectievelijk het eerste en tweede seizoen. De hoogste water productiviteit voor gewassen is gerealiseerd onder wekelijks irrigatie en varieerde 0,32 - 0,36 kg/m^3 en het laagste is verkregen bij 20 dagen irrigatie na de bloei (0,21 en 0,26 kg/m^3). Watertekort tijdens de formatie fase verminderde de opbrengst ten opzichte van volledige irrigatie, maar de vermindering was veel minder dan wanneer het tekort optrad tijdens de bloei.

De netto inkomsten bij elke irrigatie behandeling zijn onderzocht om het effect van watertekort op de zaadopbrengst economisch evalueren. De nettowinst is vooral afhankelijk van de productiekosten. Echter 8 - 9% hogere productiekosten bij de 10 dagen irrigatie intervallen vergeleken met irrigatie intervallen van 15 dagen resulteerde in een hoger netto resultaat van 721 US$/ha en 866 US$/ha in respectievelijk het eerste en tweede seizoen. Hieruit bleek dat de winst door de hogere oogst groter is dan de bijbehorende hogere irrigatie kosten voor deze behandeling.

De tweede en derde experimenten zijn uitgevoerd in respectievelijk de winter en de zomer voor het evalueren van de zaadopbrengst verkregen uit geïrrigeerde zonnebloem (10, 15 en 20 dagen irrigatie intervallen) onder twee inter-rij plant afstanden (30 en 40 cm) en daarop de cultivar Hysun 33 (Hybrid) in het winterseizoen van 2012/2013. De metingen van de zonnebloem parameters, zoals plant hoogte, diameter van de kruin, totale zaadopbrengst en 100-zaden gewicht, zijn geregistreerd. Voor de winter experimenten lieten de resultaten zien dat de hoogste zaadopbrengst werd verkregen bij een irrigatie interval van 10 dagen met 40 cm plantafstand (3290 kg/ha), gevolgd door 15 dagen interval ongeacht de inter-rij plantafstand (3120 en 3050 kg/ha) in respectievelijk het eerste en tweede seizoen. Bij beide plantafstanden is de laagste zaadopbrengst van respectievelijk 1890 en 1830 kg/ha verkregen bij een 20 dagen irrigatie interval. De hoogste water productiviteit voor gewassen van 0,45 - 0,42 kg/m^3 is behaald bij een irrigatie interval van 15 dagen en de laagste oogst en water productiviteit voor gewassen van 0,32 - 0,31 kg/m^3 zijn verkregen bij irrigatie elke 20 dagen onder de inter-rij plantafstand van 30 en 40 cm. De hoogste economische water productiviteit van 0,29 US$/m^3 is verkregen bij een irrigatie interval van 15 dagen in het eerste seizoen en de laagste economische water productiviteit van 0,17 US$/m^3 is verkregen bij een irrigatie interval van 20 dagen in het eerste seizoen. Het bleek ook dat de economische indicator een geschikt instrument kan zijn voor het beoordelen van de effecten van beperkte irrigatie en water prijzen.

De zomer experimenten zijn uitgevoerd om het effect van verschillende irrigatie intervallen vergeleken met de zaadopbrengst onder regenafhankelijk omstandigheden te bestuderen. Twee gewasvariëteiten Hysun 33 (Hybrid) en Bohooth-1 (plaatselijke variëteit) zijn geteelt respectievelijk startend in half juli 2012 en 2013. De resultaten toonden aan dat watertekort voor het gewas bij 20 dagen irrigatie intervallen in respectievelijk het eerste en tweede seizoen voor Hysun 33 de zaadopbrengst tussen 40 en 44% verminderd en voor Bohooth-1 tussen 38 en 44%. Vergroten van het irrigatie interval tot 15 dagen resulteerde echter niet in een significante vermindering van hoge zaadopbrengst (16 - 31%). De resultaten lieten zien dat de verschillen in de zaadopbrengst tussen het 10 en 15 dagen irrigatie interval niet erg significant zijn. Er zijn in het zomerseizoen tussen de twee rassen geen significante verschillen gevonden in zaadopbrengst en de water productiviteit voor gewassen onder verschillende irrigatie behandelingen en inter-rij plantafstanden. Ten opzichte van het 10 dagen interval werd bij de irrigatie intervallen van respectievelijk 15 dagen en 20 dagen de gemiddeld aangevoerde hoeveelheid water verhoogd met 6 en 11%. De bijdrage van de regen aan de wateraanvoer was echter respectievelijk 18% bij 10 dagen, 24% bij 15 dagen en 24% bij de 20 dagen intervallen. Slechts een deel van deze regen kan worden benut om te voldoen aan de waterbehoefte van het gewas in de begin fase. In dit verband geven de vergelijkingsresultaten in het per seizoen aangevoerde water tussen de drie irrigatie behandelingen (10, 15 en 20 dagen interval) aan dat het 10 dagen en het 20 dagen interval resulteren in respectievelijk een hogere en lagere irrigatie water gift.

Voor Hysun 33 is hogere water productiviteit verkregen bij irrigatie intervallen van 10 dagen (0,41 - 0,42 kg/m^3) en 15 dagen (0,39 - 0,44 kg/m^3) in het eerste en tweede seizoen, terwijl de hoogste water productiviteit (0,41 kg/m^3) voor Bohooth-1 in het eerste seizoen is verkregen bij een irrigatie interval van 10 dagen, ongeacht de plantafstand en 0,39 kg/m^3 bij hetzelfde irrigatie interval in het tweede seizoen. Tussen de twee behandelingen zijn echter grote verschillen opgetreden.

Voor alle volledige irrigatie behandelingen waren de kosten-baten verhoudingen (B/C) hoger dan 1. We kunnen bij de huidige marktprijzen dus concluderen dat de teelt van zonnebloemen in het Gezira systeem bij volledige irrigatie met 10 dagen en/of 15 dagen intervallen een economisch levensvatbare optie is. In vergelijking met de andere behandelingen zijn de totale productiekosten voor volledige irrigatie hoger, terwijl onder de heersende economische situatie het voordeel min of meer acceptabel was.

Gewas groei simulatiemodellen zijn ontwikkeld voor het voorspellen van de effecten van water en bodem op biomassa, graan opbrengst en water productiviteit van verschillende gewassen. In deze studie is het door water gestuurde gewasmodel AquaCrop, ontwikkeld door de FAO, gekalibreerd en gevalideerd voor zonnebloemen onder verschillende irrigatie regimes. De kalibratie is gedaan met behulp van gegevens verkregen uit het proefveld in 2011 en validatie met de gegevens van 2012 en 2013. De gekalibreerde en gemeten gegevens zijn vergeleken om het functioneren van het model te beoordelen. De overeenkomst tussen de gekalibreerde en gemeten waarden zijn kwantitatief geëvalueerd met behulp van de kwadratisch gemiddelde fout (RMSE) en de index van de overeenkomst (d). Hoe lager de RMSE en hoe hoger d zijn, hoe beter de modelsimulatie is.

De statistische indicatoren RMSE en d suggereren dat het model gebruikt kan worden om zeer betrouwbaar de opbrengst en de bedekkingsgraad onder volledige en het beperkte irrigatie omstandigheden te beoordelen. In dit onderzoek voorspelde het AquaCrop model de zaadopbrengst met RMSE 0,05 - 0,16 t/ha, d 0,87 - 0,98 en bedekkingsgraad met RMSE 1,9 - 10,1% en d 0.99 nauwkeuriger in vergelijking met de gesimuleerde water productiviteit voor gewassen (RMSE 0,10 - 0,14 kg/m^3 en d 0,27 - 0,64). Er was bij alle irrigatie behandelingen voor het winterseizoen een neiging tot overschatting van de water productiviteit voor gewassen. Het AquaCrop model volgde dezelfde trend in het voorspellen van de zaadopbrengst en de water productiviteit voor gewassen onder zomerse omstandigheden. De zaadopbrengsten werden voorspeld met RMSE van 0,01 - 0,12 t/ha en d 0,89 - 0,99 en het verschil tussen de voorspelde en gemeten waarden was -3.1 - 16.7%. Het model overschatte de water productiviteit voor gewassen met RMSE 0,01 - 0,03 kg/m^3 en d 0,31 - 0,88, terwijl de afwijking voor alle irrigatie behandelingen 9,8 - 37,5% was.

Op basis van deze resultaten de wordt onder omstandigheden van beperkte gegevens en opbrengstramingen onder verschillende watertoevoer scenario's toepassing van het AquaCrop model aanbevolen. Daarom is het op basis van de resultaten van deze studie raadzaam om onderzoeksprogramma's uit te voeren teneinde te komen tot beperkte irrigatie als een nieuwe irrigatie strategie met aanbevolen agronomische praktijken die leiden tot een toename in de oogst en de water productiviteit voor gewassen onder Gezira omstandigheden. Daarnaast is het raadzaam om optimale watergebruik toepassingen voor zonnebloemen te overwegen die de oogst en het rendement van irrigatie water onder verschillende klimatologische omstandigheden en veranderende markt prijzen maximaliseren.

Beperkte irrigatie is in verschillende proefvelden toegepast om het effect op de oogst van zonnebloemen en water productiviteit voor gewassen te beoordelen. De resultaten laten zien dat, behalve onder groot watertekort, de verkregen oogsten bij verschillende irrigatie niveaus hoog zijn in vergelijking met door de boeren verkregen oogsten. Bovendien was de water productiviteit voor gewassen hoger wanneer het gewas in vergelijking met volledige irrigatie minder water kreeg toegediend. Daarom bevestigen deze resultaten dat irrigatie om de 15 dagen (beperkt watertekort) aanvaardbaar is voor het verbeteren van de water productiviteit voor gewassen als er een tekort aan water in het groeiseizoen zou zijn. Als land geen beperkende factor is zou een plantafstand van 40 cm kunnen worden toegepast met irrigatie elke 10 dagen voor het maximaliseren van de opbrengst en het behalen van meer winst.

ANNEX G. About the Author

Eman R.A. Elsheikh graduated from Gezira University, Faculty of Agricultural Sciences in 2000. She works in the Agricultural Research Corporation, Land and water Research Centre, WadMedani since 2003. She started her master program in 2005 in the Water Management and Irrigation Institute, University of Gezira and completed her master study in 2007. In January 2010 she worked in the Forestry Research Centre, Khartoum until November 2011. She participated in an international short course on Integrated Watershed Management in Bangkok, Thailand in 2008. In 2009 she attended the Watershed and River Basin Management short course at UNESCO-IHE. During this course she met Prof. Bart Schultz for PhD proposal. After he had agreed, in January 2011 she got a full scholarship from the Netherlands Fellowship Programme (NFP) for her PhD study at UNESCO-IHE Institute for water education, Department of Water Science and Engineering. Her PhD research was on Water Productivity for Sunflower (*Helianthus annuus L*) under Gezira clay soil conditions, Sudan. In 2014 she gained another grant for research from the International Foundation for Sciences (IFS). She attended two annual PhD seminars during her PhD study

International short course participation during the PhD study:
1/ Sustainable management of soil and water resources, 16-27 June 2014. Izmir, Turkey;
2/ Water harvesting and small scale irrigation, 16 August - 23 September, 2013. Armidaile, Australia.

Intrenational Workshop papers:
1/ Effect of irrigation interval on yield and yield components of Sunflower in Gezira condition. 5th International conference on Sustainable Irrigation, 11 - 13 December, 2012. Adelaide, Australia.

2/ Effect of deficit irrigation on Sunflower yield in Gezira Clay conditions, Sudan. Soil Physics Properties under Climate Change, College on Soil Physics, 25 February - 1 March 2013. Trieste, Italy

3/ Improving farmer's income by optimizing water productivity. Water harvesting for small scale irrigation. Workshop. 16 - 28 February, 2014. Polokwane, South Africa.

Publications:
1/ Eman,R.H.Elsheikh, Bart Schultz, Abraham Mehari Haile and Hussein S. Adam. 2015. Effect of deficit irrigation on yield and yield components of Sunflower (*Helianthus annuus L.*) on Gezira clay soil, Sudan, Published in *African Journal for Agricultural Research*
2/ Eman, R.H. Elsheikh, Bart Schultz, Abraham Mehari Haile and Hussein S. Adam. 2015. Crop water productivity for Sunflower under different irrigation regimes and plant spacing, Gezira Scheme, Sudan, accepted by *Journal of Agriculture and Environment of International Development.*
3/ Eman, R.H. Elsheikh, Bart Schultz, Abraham Mehari Haile and Hussein S. Adam. Calibration and validation of AquaCrop Model to simulate crop water productivity for Sunflower under different irrigation management in Gezira Scheme, Sudan. Submitted.

Netherlands Research School for the
Socio-Economic and Natural Sciences of the Environment

D I P L O M A

For specialised PhD training

The Netherlands Research School for the
Socio-Economic and Natural Sciences of the Environment
(SENSE) declares that

Eman Rahamtalla Ahmed Elsheikh

born on 3 June 1974 in Khartoum, Sudan

has successfully fulfilled all requirements of the
Educational Programme of SENSE.

Delft, 8 October 2015

the Chairman of the SENSE board

Prof. dr. Huub Rijnaarts

the SENSE Director of Education

Dr. Ad van Dommelen

The SENSE Research School has been accredited by the Royal Netherlands Academy of Arts and Sciences (KNAW)

K O N I N K L I J K E   N E D E R L A N D S E
A K A D E M I E   V A N   W E T E N S C H A P P E N

The SENSE Research School declares that **Ms Eman Elsheikh** has successfully fulfilled all requirements of the Educational PhD Programme of SENSE with a work load of 50.6 EC, including the following activities:

SENSE PhD Courses

o   Environmental Research in Context (2011)
o   Research in Context Activity: 'Organising two field days with farmers and engineers, to discuss how water may be used efficiently during critical crop stages and recommended irrigation scheduling' (2013)

Other PhD and Advanced MSc Courses

o   Aspect of Irrigation and Drainage, UNESCO-IHE, Delft (2011)
o   How to write a scientific paper, Wageningen University (2011)
o   Soil physical properties and processes under climate change, Trieste, Italy (2013)
o   Training in measurements of irrigation, soil moisture content and training in irrigation design, UNESCO-IHE, Delft (2011)

External training at a foreign research institute

o   Water harvesting and small scale irrigation, University of New England, Australia (2013)
o   Sustainable Management of Soil and Water Resources, International Agricultural Research and Training Center, Izmir, Turkey (2014)

Management and Didactic Skills Training

o   Organising field Day in agricultural research corporation, Sudan (2011)

Oral Presentations

o   *Effect of irrigation intervals on yield and yield components of sunflower in Gezira conditions.* International Conference for Irrigation and Drainage, 11-13 December 2012, Adelaide, Australia
o   *Effect of water deficit on sunflower yield under Gezira clay soil.* Workshop 'Soil Physics Properties under Climate Change', 25 February-1 March 2013, Trieste, Italy
o   *Improving farmer's income by optimizing water productivity.* International Workshop 'Water harvesting and small scale irrigation', 16-28 February 2014, Polokwane, South Africa
o   *Crop water productivity for sunflower under different irrigation intervals and two plant spacing at Gezira Conditions.* Annual PhD Seminar at UNESCO-IHE, 20-21 September 2014, Delft, The Netherlands

SENSE Coordinator PhD Education

Dr. ing. Monique Gulickx

Printed and bound by CPI Group (UK) Ltd, Croydon, CR0 4YY

21/10/2024

01777101-0004